Common
Legumes
of the
Great Plains

"Thorough, well-researched. . . . A very useful book for a variety of people in natural resource fields."—Gary E. Larson, South Dakota State University.

Woolly locoweed, hog peanut, black medic, and false indigo are popular names for only a few of the 180 species of legumes to be found on the Great Plains. Second only to the grasses in abundance and economic value, legumes are important as dietary staples for humans and animals, as soil erosion controls, fertilizers, firewood, medicines, and landscape plants. Because they have so many uses, and because some are poisonous, it is important to be able to identify them. James Stubbendieck and Elverne C. Conard have produced the first illustrated guide to the legumes to be found in thirteen central plains states and three Canadian provinces.

Common Legumes of the Great Plains presents keys to the families, genera, and species of these plants, describing the life span, origin, and height of each, as well as their stems, leaves, flowers, and fruit. Included are distribution maps and botanical illustrations for the 106 most common and important species. Relative habitat and abundance are discussed, and so are seed production, means of establishment, landscape value, and erosion control characteristics. Based on modern taxonomy, the book contains an index and glossary. It is certain to become a standard reference of value to the rancher, farmer, and naturalist as well as to the botanist and taxonomist.

James Stubbendieck is a professor of agronomy and Elverne C. Conard is an associate professor emeritus of agronomy at the University of Nebraska–Lincoln.

About the authors:

James Stubbendieck is a professor of agronomy (range ecology) at the University of Nebraska–Lincoln. A native of Nebraska, he received his B.S. and M.S. degrees at the University of Nebraska and his Ph.D. at Texas A&M University.

Elverne C. Conard is an associate professor emeritus at the University of Nebraska–Lincoln. He was born in Illinois and grew up in Wyoming. He received his B.S. degree from Colorado A&M College, his M.S. from the University of Nebraska, and his Ph.D. from Texas A&M University.

Common Legumes of the Great Plains

An Illustrated Guide

By James Stubbendieck
and Elverne C. Conard

Illustrated by

Bellamy Parks Jansen

University of Nebraska Press
Lincoln and London

*Library of Congress Cataloging-in-
Publication Data*
Stubbendieck, James L.
    Common legumes of the Great Plains:
an illustrated guide / by James Stubben-
dieck and Elverne C. Conard; illustrated
by Bellamy Parks Jansen.
        p.   cm.
    Bibliography: p.
Includes index.
ISBN 0-8032-4204-2 (alk. paper)
    1. Leguminosae—Great Plains—
Identification. 2. Leguminosae—Great
Plains—Pictorial works. I. Conard,
Elverne C., 1909–   . II. Jansen, Be-
llamy Parks. III. Title.
QK495.L52S92   1989
583©.320978.dc19   CIP 88-27690

In memory of Esther

CONTENTS

PREFACE

I was fortunate to begin my teaching and research career in the Department of Agronomy, at the University of Nebraska-Lincoln in 1978 as successor to Dr. Elverne C. Conard. Dr. Conard had retired from the range ecology position, and I felt honored to be his replacement and teach the courses I had taken from him ten years previously. While exploring files in the range plant herbarium, I discovered a key to the legumes of Nebraska that Dr. Conard had written and used in his classroom. It seemed to be an excellent beginning to a book. But a problem with legume identification is the similarity in appearance of many species. Any book on the topic would have to be illustrated, but photographs seldom show enough detail, and so the project was abandoned before it had a chance to be started.

In 1981, I initiated work on North American Range Plants (University of Nebraska Press, 3rd edition, 1986) and needed a few illustrations to complete the book. Dr. Ronald R. Weedon, Department of Biology, Chadron State College, introduced me to Bellamy Parks, a student and biological illustrator at Chadron State College. She completed those illustrations. Later, she moved to the University of Nebraska, married, and was looking for a project requiring her excellent illustration talent.

The team was then complete—Dr. Conard to select material for illustration and to contribute to the writing, Bellamy Parks Jansen to illustrate the legumes, and me to coordinate and write. The project was expanded to include the Great Plains rather than Nebraska only.

We drew on information from numerous publications during the preparation of this book and examined thousands of herbarium specimens and living plants. We express thanks and appreciation to Dr. Margaret R. Bolick, Curator of Botany and Associate Professor, Nebraska Museum and School of Biological Sciences, University of Nebraska-Lincoln, and Dr. Ronald R. Weedon, Curator of Botany and Professor, Department of Biology, Chadron State College, Chadron, Nebraska, for loaning herbarium specimens for examination.

We also acknowledge Dr. Lowell E. Moser, Dr. Steven S. Waller, Dr. Francis A. Haskins, Department of Agronomy, University of Nebraska-Lincoln, and Dr. James T. Nichols, University of Nebraska West Central Research and Extension Center, North Platte, for answering specific technical questions as well as being willing to take time to listen and discuss problems we encountered during the preparation of the manuscript. We are grateful to Chuck Butterfield and Theresa Flessner for their repeated reviews of the manuscript as well as for contributing to the preparation of distribution maps and to the glossary. Cindy Veys, Lincoln, Nebraska, prepared Figure 1 and the base map for plant distributions. We especially thank Dr. Gary E. Larson, Department of Biology, South Dakota State University, for critically reviewing the manuscript. Pat Knapp, University of Nebraska Press, is acknowledged for her helpful suggestions and encouragement. Finally, we gratefully acknowledge Sue Peterson and Charlene Cunningham for typing the manuscript, suggesting improvements in the format, and correcting the errors. Without their dedicated help, this book could not have been completed.

This book was written to improve the reader's background, knowledge, and appreciation of leguminous plants. We believe individuals with an extensive botanical knowledge, as well as those with a passing interest in plants, will gain from this book. Naturalists, students, reclamation specialists, land managers, and livestock producers will find valuable information and answers to many of their questions in the following pages.

James Stubbendieck

INTRODUCTION

The Great Plains

The Great Plains region occupies a vast area in central North America (Figure 1). It comprises all or part of 13 states of the United States and 3 Canadian provinces and has an area of more than 500 million acres (200 million hectares). It measures nearly 2,000 miles (3,300 kilometers) from north to south and more than 900 miles (1,500 kilometers) from west to east.

The continental climate of the Great Plains is one of extremes. Temperatures commonly vary from −20° F (-29° C) in winter to 110° F (43° C) in summer. Winds are strong in winter and spring. Precipitation varies from about 10 inches (250 millimeters) in the west to more than 40 inches (1,000 millimeters) in the east because of the rain shadow caused by the Rocky Mountains which form the western boundary of the Great Plains. It has a continental type climate where much of the precipitation falls during the growing season. Drought is common.

Soils of the Great Plains are highly variable in texture and fertility. The loess soils of the central and eastern regions are deep and fertile. Most of these areas are cultivated. The Nebraska Sandhills occupy about 12 million acres (5 million hectares), and smaller areas of sandy soils are found in other states. Rocky soils are common in the Kansas Flint Hills and in the northern states and Canada. The soils are generally calcareous in the west and slightly acidic in the east.

The Great Plains before settlement was primarily prairie with a few relatively small areas of forest. Prairie vegetation is composed of grasses, grasslike plants, shrubs, and forbs (including legumes). The tallgrass or true prairie occupied the eastern areas receiving the larger amounts of precipitation; the shortgrass prairie the western areas with the lowest amounts of precipitation; and the mixed prairie the central region.

The natural vegetation, including native legumes, evolved to grow and reproduce readily in these extremes of climate, variable soils, frequent prairie fires, and competition from other plants. In addition, it developed ways of withstanding herbivores varying in size from insects to large ungulates.

Abundance and Value of Legumes

Legumes are one of the largest and most important groups of plants in the Great Plains. They are second to grasses (Poaceae) in economic value and third in abundance following grasses and composites (Asteraceae). More than 40 genera and 180 species are found on the Great Plains. Some are rare, while others are found in virtually every county.

Legumes are important in the diets of humans in many parts of the world. Numerous kinds of peas and beans are harvested green or upon maturity. This book does not include the legumes eaten by humans, but nearly all of the legumes included here are consumed by domestic livestock and/or wildlife. The most important hay crops are alfalfa and clovers. However, certain species of legumes are among the most poisonous

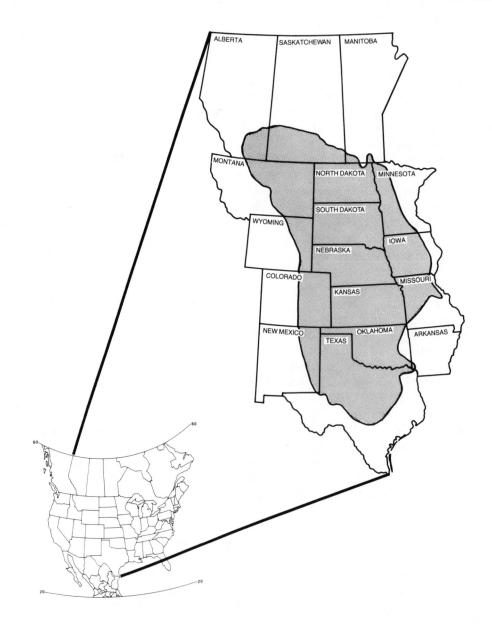

Figure 1
The Great Plains

Figure 2
The papilionaceous flower from different perspectives

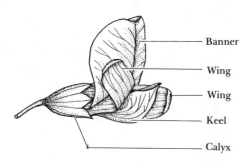

Banner

Wing

Wing

Keel

Calyx

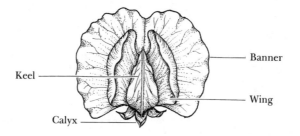

Keel

Banner

Wing

Calyx

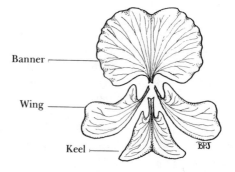

Banner

Wing

Keel

plants found in nature. Some can be managed to be valuable forage producers but are poisonous when managed improperly.

Some legumes are important for prevention and control of soil erosion. The foliage intercepts the raindrops and breaks their force. Roots bind the soil and foliage covers it, protecting it from the erosive forces of both wind and water.

Most legumes improve soil fertility through a symbiotic relationship with certain bacteria (*Rhizobium* spp.). Individual species of *Rhizobium* are usually specific for a single species of legume. The rhizobia form nodules on the roots of legumes. The legumes furnish the necessary food for energy that enables the rhizobia to change nitrogen from the atmosphere into a form that plants can use. This process is called nitrogen fixation. The surplus nitrogen improves soil fertility and is used by other plants.

Many legumes have an aesthetic value. The flowers of some species are large and showy, whereas those of others are small and delicate. The flower and foliage color and form of growth are varied, making legumes valuable species for landscaping.

The legume trees furnish excellent firewood, and many species are important for honey production. Food, medicinal, and other uses of legumes by American Indians and pioneers are described in this book.

Keys

Keys to the families, genera, and species are based on floral and/or vegetative features. In some instances fruit characteristics have been used when the other characteristics are not particularly diagnostic. Keys were constructed using both living and pressed material. Most of the keys are new, but a few were previously published and have been extensively modified to fit material from the Great Plains.

The keys are artificial. No attempt was made to key closely related genera or species together. Dichotomous choices are identified by the same capital letters.

Taxonomy and Common Names

Nomenclature generally follows that of the *Flora of the Great Plains* (1986) and *A Synonymized Checklist of Vascular Flora of the United States, Canada, and Greenland*, Volume II (1980). More than 120 references on all facets of legumes are found in the Selected References beginning on page 307. The term "legume" is generally used in association with members of the Fabaceae (Bean Family). For purposes of simplicity and because of similarity of species, the term "legume", as used in this text, refers collectively to all members of the Mimosaceae (Mimosa Family), Caesalpiniaceae (Caesalpinia Family), and Fabaceae (Bean Family).

Synonymy has been included in the text and the index. Some relatively recent changes, such as the inclusion of species formerly in *Petalostemon* in *Dalea*, may be difficult for individuals with a knowledge of scientific names. Extensive use of synonyms throughout the text may reduce the difficulty. Differences of opinion with respect to nomenclature will probably continue.

The meanings of scientific names have been included. Where necessary, the names have been divided and the meanings are given for each part. Abbreviations given for the source of the names are Gk. (Greek), Lat. (Latin), Tr. (Tartar), and Ar. (Arabic). Names derived from other sources also are described.

Authorities for scientific names, including synonyms, have been included in abbrevi-

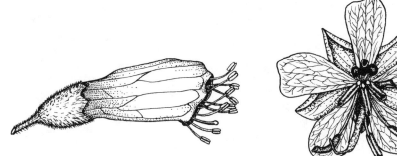

Figure 3
Other legume flowers

FABACEAE (*Amorpha*)

CAESALPINIACEAE (*Cassia*)

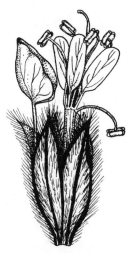

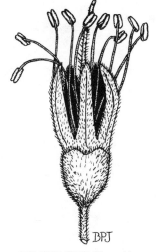

FABACEAE (*Dalea*)

MIMOSACEAE (*Prosopis*)

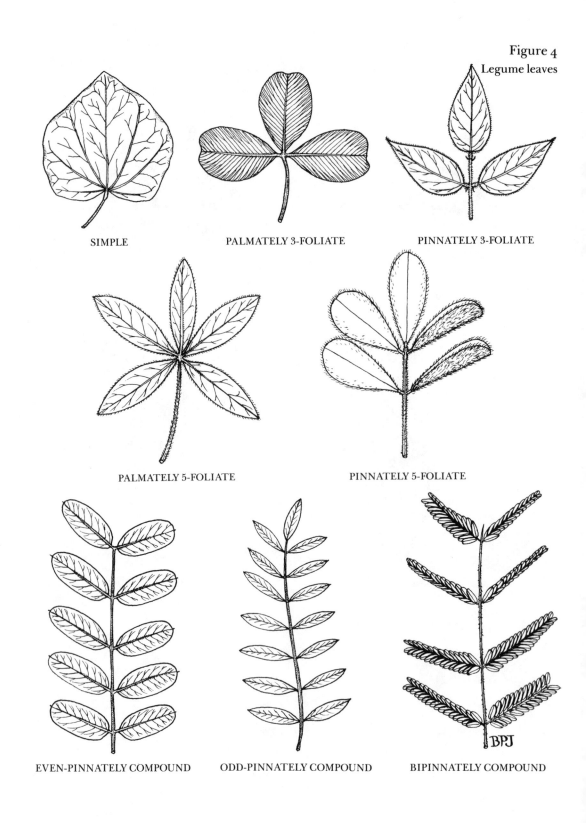

Figure 4
Legume leaves

SIMPLE

PALMATELY 3-FOLIATE

PINNATELY 3-FOLIATE

PALMATELY 5-FOLIATE

PINNATELY 5-FOLIATE

EVEN-PINNATELY COMPOUND

ODD-PINNATELY COMPOUND

BIPINNATELY COMPOUND

ated form. Complete names, life spans, and positions are found in the section entitled Authorities starting on page 301.

A single common name has been selected for each species on the basis of frequency of use. Common names have been restricted to two words and hyphens have been eliminated. Other common names used in the literature are included. The selected common name and other common names of each species are included in the index.

Similar species within a genus are briefly described. These species were not selected for full descriptions because of relatively small numbers or a restricted range. Differences between these additional species and those given full treatment are discussed.

Illustrations

Complete illustrations of 107 species have been included. These detailed pen and ink drawings illustrate the plant in flower and in fruit. Individual features, such as the flower and leaf, are often greatly enlarged and shown in detail. All illustrations were prepared by Bellamy Parks Jansen from living plant material and/or from herbarium specimens. These illustrations present the average appearance of the numerous specimens examined rather than that of a single specimen. Size of all portions of the illustration are given in the figure captions.

Maps

Distribution maps were compiled from herbarium specimens and reference materials. They illustrate the range in which the species is mostly likely to occur. The species may occasionally occur outside of that range, and it is not equally abundant throughout the range.

Descriptions of Individual Species

Descriptions of the families, genera, and species are provided. Descriptions of stems, leaves, inflorescences, flowers, and fruit follow the descriptions of the life span, origin, and height of each species. Ranges of measurements of plant parts are provided. These measurements will include at least 90 percent of the plants encountered, but extreme environmental conditions may cause the plant part to be either larger or smaller. Technical terms are defined in the Glossary starting on page 295.

Chromosome numbers are listed at the end of each technical description. These numbers were taken from the literature, and no chromosome numbers were determined specifically for this book.

Following listings of synonyms and other common names, relative abundance and habitat are discussed. Range of flowering interval is presented. Relative palatability and value to domestic livestock and wildlife are given. Responses to grazing, relative importance, and poisonous principles are included where applicable. Seed production, means of establishment, and landscape and/or value for erosion control are discussed. The final section may include a discussion of similar species.

Characteristics of Legumes

Most legumes have papilionaceous flowers (Figure 2). These irregular flowers are made up of five petals. The banner is the single upper petal. It is sometimes called the stan

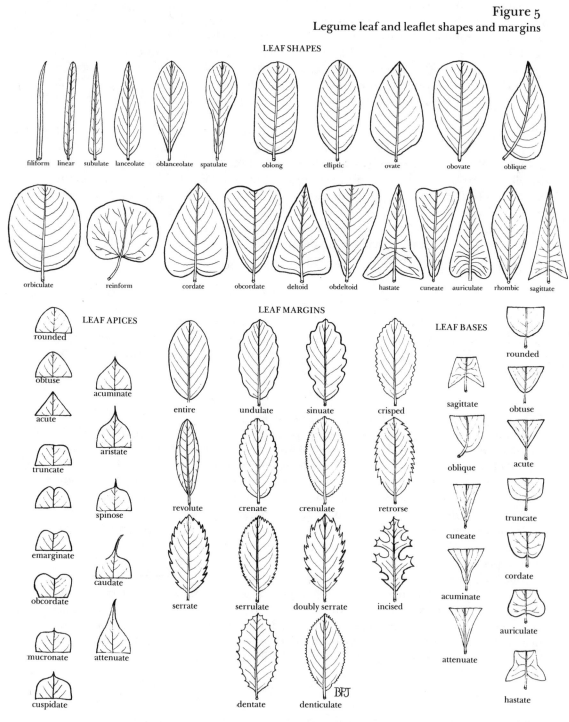

Figure 5
Legume leaf and leaflet shapes and margins

LEAF SHAPES

filiform linear subulate lanceolate oblanceolate spatulate oblong elliptic ovate obovate oblique

orbiculate reinform cordate obcordate deltoid obdeltoid hastate cuneate auriculate rhombic sagittate

LEAF APICES

rounded
obtuse
acute acuminate
truncate aristate
emarginate spinose
obcordate caudate
mucronate attenuate
cuspidate

LEAF MARGINS

entire undulate sinuate crisped

revolute crenate crenulate retrorse

serrate serrulate doubly serrate incised

dentate denticulate

BPJ

LEAF BASES

sagittate rounded
oblique obtuse
cuneate acute
acuminate truncate
attenuate cordate
auriculate
hastate

Figure 6
Legume fruits

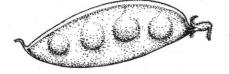

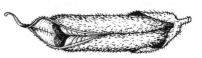

SEVERAL-SEEDED LEGUME   BILOCULAR LEGUME

ONE-SEEDED LEGUME   COILED LEGUME   INFLATED LEGUME

DEHISCENT LEGUME   INDEHISCENT LEGUME   COILED AFTER DEHISCENCE

BPJ

TWO-JOINTED LOMENT   FOUR-JOINTED LOMENT

dard. The two lateral petals are the wings, and the two lower petals are the keel. The keel petals are occasionally fused.

Other types of legume flowers are shown in Figure 3. The *Amorpha* flower has a single petal. The *Cassia* flower has five nearly regular petals. In the *Dalea* flower, the petals are modified stamens. The *Prosopis* flower is very small and is nearly regular with five petals. Leaf arrangements are presented in Figure 4. Most legumes have compound leaves with leaflets arranged either palmately or pinnately. Examples of odd- and even-pinnately compound leaves and bipinnately compound leaves are illustrated.

Leaf characteristics are important for the identification of legumes. Leaf shapes, apices, bases, and margins are illustrated in Figure 5. This figure is a comprehensive illustration of these characteristics, and not all are represented in the legumes described in this book.

The fruit of the leguminous plants is either a legume, sometimes improperly called a seed pod, or a loment. Fruit characteristics are necessary for identification of some species. Several types of legumes and two types of loments are illustrated in Figure 6.

TAXONOMY OF LEGUMES

The group of plants comprising the legumes have been variously treated. Conventionally, most taxonomists included all legumes in one family, the Fabaceae or Leguminosae. The large family was subdivided into the three subfamilies Mimosoideae, Caesalpinioideae, and Papilionoideae. This division was made primarily on the basis of floral differences. Other taxonomists have elevated these subfamilies to full family ranking in accordance with the International Legume Conference held at the Royal Botanic Gardens, Kew, England, in 1978. As a result, most authors now recognize three families of legumes: Mimosaceae (Mimosa Family), Caesalpiniaceae (Caesalpina Family), and Fabaceae (Bean Family).

Key to the Families

A. Flowers regular or nearly so, small, in dense heads or racemes; petals 4 or 5, inconspicuous, valvate in bud; stamens strongly exserted; leaves bipinnately compound
....I. **Mimosaceae**

A. Flowers more or less irregular, sometimes imperfectly papilionaceous; petals imbricated in the bud; leaves often other than bipinnately compound.

    B. Corolla irregular, imperfectly papilionaceous, or not at all irregular; banner enclosed by the lateral petals in the bud; leaves bipinnately compound, pinnately compound, or simple.................................II. **Caesalpiniaceae**

    B. Corolla usually papilionaceous (mostly 5 petals, 1 in *Amorpha* and 1 plus 4 petal-like staminodes in some *Dalea* spp.) with the keel petals more or less united; banner enclosing the lateral petals in bud; leaves mostly palmately 3-foliate or pinnately compound, never bipinnately compound................III. **Fabaceae**

FAMILY I. MIMOSACEAE R. BR.

Mostly trees, shrubs, woody vines, or more rarely herbs; stems armed or unarmed; leaves alternate, usually even-bipinnately compound or less commonly pinnately compound; pinnae and leaflets few to many, leaflets sometimes touch-sensitive; petiole and rachis bearing 1 to several glands, depressed or stalked; stipules conspicuous or inconspicuous, sometimes reduced to spines; inflorescences of densely-flowered cylindric spikes, umbels, globose heads, or rarely racemes, pedunculate from axils; flowers regular, perfect or rarely unisexual, usually small; calyx usually gamosepalous, 5-lobed, lobes minute; corolla gamopetalous or polypetalous, valvate in bud, 5 petals (rarely 4); stamens usually 10 or multiples thereof, sometimes 5 (rarely 1), distinct or united basally, filaments usually long- exserted; legume straight, curved, or spirally twisted, usually 2-valved, dehiscent, occasionally transversely separated between the seeds.

A family of about 40 genera and at least 2,000 species. They are chiefly found in the tropics and warm temperate areas. The family is represented in the Great Plains by 6 genera. Three (*Acacia, Mimosa,* and *Neptunia*) are not common.

Key to the Genera

A. Shrubs or small trees... 2. *Prosopis*
A. Herbs
    B. Stems and legumes armed with recurved prickles; legumes long and slender; flowers pink; petals united to about the middle; lowest pair of pinnae without a gland between them .... 3. *Schrankia*
    B. Stems and legumes smooth, unarmed; legumes short, flat, curved, in relatively compact globose heads; flowers white to greenish-white; petals free; lowest pair of pinnae with a gland between them.......................... 1. *Desmanthus*

1. DESMANTHUS Willd.

[*desme* (Gk.): a bundle; + *anthos* (Gk.): flower, referring to the dense heads.]

Perennial herbs or shrubs, erect or spreading, bipinnately compound leaves; leaflets small and numerous; flowering heads axillary, long-peduncled; petals 5, white to greenish-white; calyx gamosepalous, 5-lobed, tube campanulate; stamens 5 (or 10), usually long-exserted; legumes oblong to linear, straight or curved, few- to several-seeded.

Thirty species have been described in tropical and subtropical America. One species is common in the Great Plains; 2 others are occasionally collected.

Figure 7   *Desmanthus illinoensis*

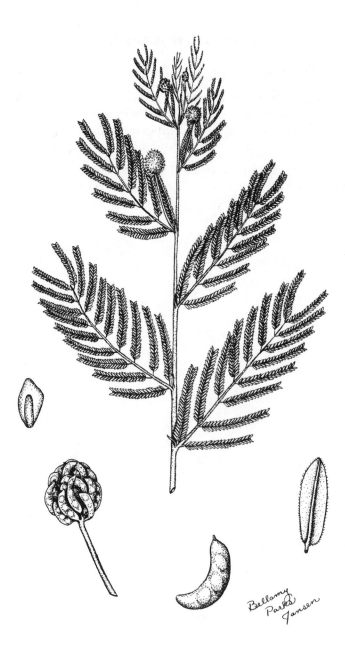

Desmanthus illinoensis (Michx.) MacM.   Illinois bundleflower (Figure 7)

[*illinoensis*: of or from Illinois.]

Life Span: perennial. *Origin*: native. *Height*: (3)8–20 dm. *Stems*: herbaceous, erect or ascending, slightly angled (grooved), glabrous or nearly so, from a somewhat woody caudex. *Leaves*: alternate, bipinnately compound, 5–10 cm long; pinnae 6–12(16) pairs, 2–4 cm long, gland present between pinnae pairs or only between the lower pair; leaflets 15–30 pairs, 2–5 mm long, linear to oblong, glabrous to ciliate, midvein prominent; stipules setaceous to filiform, 4–10 mm long, usually persistent. *Inflorescences*: axillary heads, many-flowered; peduncles ascending, 2–6 cm long. *Flowers*: petals 5, united to middle, becoming separate, 2 mm long, white to whitish-green, in a globose head; calyx tube campanulate, 5-lobed, 1 mm long; stamens 5, usually long-exserted. *Fruit*: legume, thin, strongly curved, 1–2.5 cm long, 4–7 mm wide, reddish-brown, numerous in a globose head, glabrous, few- to several-seeded; seeds variable, rhombic, 3–5 mm long, nearly as wide, yellowish-red to brown. 2n=28.

Synonyms: *Acuan illinoensis* (Michx.) O. Ktze., *Mimosa illinoensis* Michx.
Other Common Names: false sensitiveplant, Illinois desmanthus

Illinois bundleflower is common in dry or moist soil of prairies, open wooded slopes, ravines, waste places, and roadsides. It flowers from June to August. It is readily eaten by all classes of livestock, deer, and pronghorn. It decreases with heavy grazing. It is one of the most important native prairie legumes.

Illinois bundleflower produces abundant seeds. Seeds are readily consumed by birds and rodents. It is easily established from seed, and seed is commercially available. Illinois bundleflower is occasionally seeded with grasses for revegetation. It has moderate value for erosion control and limited value for landscaping. Its leaves are very weakly touch-sensitive and infold in strong sunlight.

Desmanthus cooleyi (Eat.) Trel., Cooley desmanthus, and *D. leptolobus* T. & G., slenderlobed bundleflower, are occasionally found in the Great Plains. Legumes of these species are linear. *Desmanthus cooleyi* has small stipules (2 mm long or less) and rhombic seeds and is found in the extreme southwestern portion of the Great Plains. *Desmanthus leptolobus* has stipules 4–6 mm long and obovate seeds. It is most commonly found in the southern portion of the Great Plains.

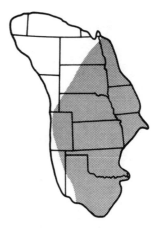

2. PROSOPIS L.

[*prosopis* (Gk.): an ancient name of an unidentified plant.]

Low woody shrubs or small trees; stems usually armed with straight, stout spines; leaves bipinnately compound with 1 to several pairs of pinnae; leaflets many, narrow; inflorescences cylindric spikes, spikelike racemes, or globose heads; flowers yellow or yellowish-brown; calyx 5-lobed, usually small; petals 5; stamens 10, anthers bearing apical glands; legumes long, nearly straight, usually constricted between seeds.

A genus of about 40 species in drier, subtropical regions of North and South America, Africa, and Asia. One species occurs in the southern Great Plains.

Figure 8 *Prosopis glandulosa*

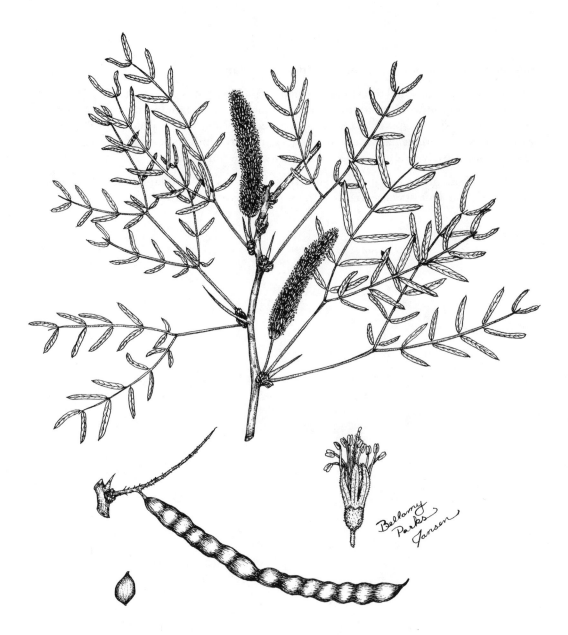

Prosopis glandulosa Torr.   Honey mesquite (Figure 8)

[*glans* (Lat.): a gland, in reference to its numerous glands.]

Life Span: perennial. **Origin**: native. **Height and Form**: shrub or small tree, with a rounded crown, 1–4(6) m tall. **Twigs**: rigid, glabrous, reddish-brown or grayish-brown, zig-zag, much branched, usually armed with 1 or 2 stout spines at the nodes. **Trunks**: single or multiple. **Leaves**: deciduous, alternate, even-bipinnately compound, usually 2 pinnae per leaf; pinnae 6–15 cm long, each with 12–30(36) leaflets; leaflets linear to oblong, terminal pair usually curved, 2–6 cm long, 2–3 mm wide, glabrous or nearly so, sessile or nearly so, margin entire, apex acute and mucronate, base obtuse, lateral veins indistinct; petiole 5–8 cm long, with a circular gland on upperside at the base of the pinnae; stipules modified into spines (rarely spineless), to 5 cm long, rigid, straight. **Inflorescences**: axillary spikelike racemes, pendulous, 7–9 cm long; peduncle glabrous, 1–3 cm long with 1–3 caducous bracts. **Flowers**: perfect, greenish-yellow, fragrant; calyx tube campanulate, 1 mm long, 5-lobed, the lobes triangular, 0.3–0.4 mm long, finely pubescent at the tips; petals 5, elliptic to obovate, 3 mm long, distinct, pubescent within; stamens 10; pedicels glandular, 0.5 mm long. **Fruit**: legumes in clusters of 2 or 3, linear, straight or nearly so, 10–15(20) cm long, 1 cm wide, slightly constricted between the seeds, indehiscent; seeds 2–10, ovoid, flattened, tan to brown, 6–7 mm long, 4.5 mm wide. $2n=56$.

Synonyms: *Neltuma glandulosa* (Torr.) Britt. & Rose, *Prosopis chilensis* var. *glandulosa* (Torr.) Standl., *P. juliflora* (Sw.) DC., *P. odorata* Torr. & Frem.
Other Common Names: mesquite, glandular mesquite, mezquite

Honey mesquite is common on dry sandy or gravelly open prairie. It is abundant and a serious weedy species on abused rangeland. It increased following the control of fires. It flowers from May to July. It furnishes poor to good forage for domestic livestock and deer. Flowers are an important honey source. Seeds are important to many species of wildlife. They serve as emergency feed for livestock. Ingestion of large quantities of foliage may cause rumen stasis and impaction.

Honey mesquite wood is used for fuel, railroad ties, and furniture. The seeds were important in the diets of native Americans. They pounded the seeds into flour and brewed beer from them.

Honey mesquite is occasionally used for an ornamental, especially the spineless selections. It is a good soil binder. Honey mesquite is the most widespread and serious weedy shrub in the extreme southern Great Plains.

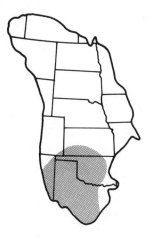

3. SCHRANKIA Willd.

[*Schrankia*: named for Franz von Paula von Schrank (1747–1835), German botanist.]

Perennial herbs (or shrubs in the south), armed with recurved prickles or hooked thorns; leaves bipinnately compound, sensitive; leaflets small; flowers numerous in axillary and peduncled heads, perfect or unisexual; calyx minute; corolla funnelform, 5-lobed, pink to purple, gamopetalous; stamens commonly 10, occasionally 8 to 13, filaments elongate, slender; legume linear to oblong, 4-valved, strongly ribbed, usually prickly, several-seeded.

About 25 species are present in tropical and warm temperate America. One is common in the Great Plains, and a second appears in the southwestern portion of the area.

Figure 9   *Schrankia nuttallii*

Schrankia nuttallii (DC.) Standl.   Sensitive brier (Figure 9)

[*nuttallii*: named after Thomas Nuttall (1786–1859), plant explorer and naturalist.]

Life Span: perennial. ***Origin***: native. ***Height***: prostrate or spreading, 0.5–1(2) m long. ***Stems***: herbaceous, strongly ribbed, armed with recurved prickles, glabrous. ***Leaves***: alternate, bipinnately compound, 6–15 cm long; pinnae 4–8 pairs, 2–5 cm long; leaflets 9–15 pairs, glabrous, oblong or elliptic, 3–9 mm long, 1–1.5 mm wide, veins prominent, reticulate beneath, apex sharply pointed; stipules persistent, linear-lanceolate, 3–6 mm long, glabrous. ***Inflorescences***: axillary heads, dense, globose, 1.5–3 cm in diameter, many-flowered; peduncles 3–9 cm long, prickly. ***Flowers***: pink (rarely rose or purple), sessile; calyx minute; corolla funnelform, 2.5–4 mm long, lobed at apex; stamens (8)10–12 with elongate slender filaments, distinct or united only at their base. ***Fruit***: legume linear, 4-valved, 4–12 cm long, to 7 mm wide, strongly ribbed, densely prickly on the ribs, not flattened, golden-brown, many-seeded; seeds brown to black, smooth, slightly flattened, 3–5 mm long, variable in shape (generally ovate to rhombic). 2n=26.

Synonyms: *Leptoglottis nuttallii* DC. *ex* Britt. & Rose, *Schrankia uncinata* Willd.
Other Common Names: bashful brier, cat-claw sensitivebrier, sensitive rose, shame vine

Sensitive brier grows on sterile, dry hill-sides, in woodlands, and in ravines. It is most common on sandy and gravelly soils. It flowers from April to September. The leaves fold and droop when touched. Domestic livestock graze the tender branches before the prickles harden. Deer and wild turkeys eat the leaves. Upland birds and rodents eat the seeds. Seed production is low due to destruction by insect larvae.

Seeds are seldom commercially available. Plants are sometimes sold as a curiosity because of the touch response. Sensitive brier has no value for erosion control or landscaping. Seeds have been used for laxatives.

Schrankia occidentalis (Woot. & Standl.) Standl., western sensitivebrier, is found in the southwestern portion of the Great Plains and is similar to *Schrankia nuttallii*. Only the midrib of the leaflet is prominent in western sensitivebrier. In addition, its seeds are quadrate and over 5 mm long, and the legumes are usually puberulent as well as prickly.

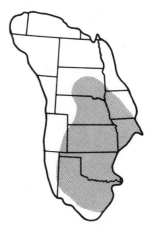

FAMILY II. CAESALPINIACEAE R. BR.

Trees, shrubs, or perennial (rarely annual) herbs; leaves alternate, pinnately compound, bipinnately compound, or rarely simple, sometimes with translucent dots, alternate, usually with stipules and lacking stipels; inflorescences usually showy racemes or panicles; flowers usually white or yellow (rarely pink or red), perfect or unisexual, irregular or nearly regular; calyx polysepalous or gamosepalous, 5-lobed, imbricate or separate; petals usually 5, sometimes rudimentary or absent, imbricate, never valvate, upper petal distinctive and innermost in bud; stamens 5 to 10 (rarely fewer), mostly free or joined; ovary sessile; legume sometimes indehiscent, sometimes with winged sutures, sometimes fleshy; seeds usually with abundant endosperm.

The family contains up to 135 genera and about 2,200 species. They occur chiefly in tropical and semi-tropical areas. It is represented in the Great Plains by 6 genera. Five are included in the key. The sixth is *Caesalpinia* with a single species that is scattered in the southwestern Great Plains.

Key to the Genera

A. Trees or tall shrubs
- B. Leaves simple, entire, cordate; flowers imperfectly papilionaceous, appearing before the leaves, rose or pink . . . . 2. *Cercis*
- B. Leaves compound; flowers not papilionaceous, appearing with or after the leaves, greenish-white to pinkish-white
  - C. Stems unarmed; flowers in terminal racemes or panicles; leaflets 4 cm long or more, entire; legumes thick and woody . . . . 4. *Gymnocladus*
  - C. Stems usually armed; flowers in axillary spikelike racemes; leaflets 4 cm long or less, crenulate; legumes large, thin, flat, coriaceous . . . . 3. *Gleditsia*

A. Herbs
- D. Leaves bipinnately compound . . . . 5. *Hoffmanseggia*
- D. Leaves pinnately compound . . . . 1. *Cassia*

1. CASSIA L.

[*kassia* (Gk.): ancient name for some plant.]

Perennial or annual herbs; leaves even-pinnately compound, leaflets numerous, entire; petiole bearing a gland between or below leaflets; stipules persistent or caducous; flowers in terminal or axillary racemes; petals equal or 1 larger than the others; stamens 10, trimorphic; ovary pubescent; fruits flattened, erect or pendant, many-seeded.

Cassia is a large genus of several hundred species of mostly subtropical to tropical trees, shrubs, and herbs. Six herbaceous species are found in the Great Plains. Two are common.

A. Leaflets 9–20 mm long; stipules persistent; fruit usually 4–6 cm long; plant annual
. . . . .1. *C. chamaecrista*

A. Leaflets 2–8 cm long; stipules caducous; fruit usually 6.5–10 cm long; plant perennial . . . . .2. *C. marilandica*

Figure 10   *Cassia chamaecrista*

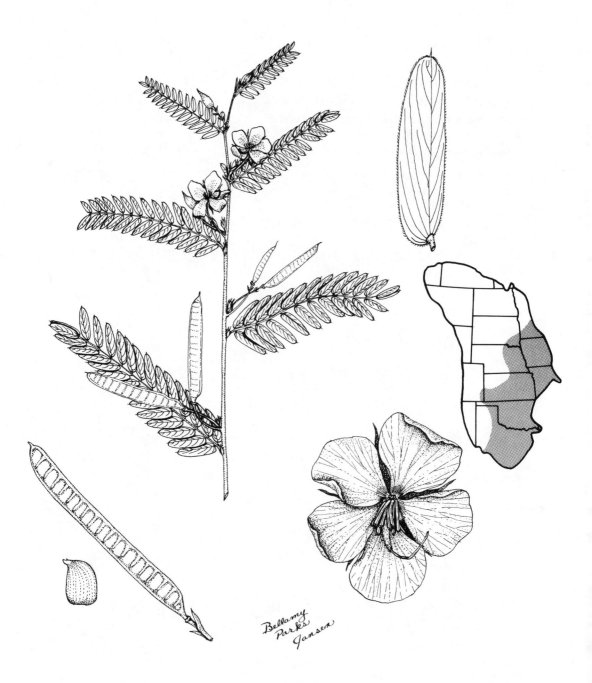

1. *Cassia chamaecrista* L.   Showy partridgepea (Figure 10)

[*chamai* (Gk.): dwarf; + *crista* (Lat.): a crest; in reference to the spot at the base of 4 of the petals.]

Life Span: annual. *Origin*: native. *Height*: 1.5–9(12) dm. *Stems*: herbaceous, erect or ascending, branching freely from the base, glabrous to minutely pubescent. *Leaves*: alternate, even-pinnately compound, 3–11 cm long; leaflets 12–36, oblong, (4)9–20 mm long, 2–4.5 mm wide, asymmetrical, sparsely hairy on margins, apex obtuse, mucronate and base oblique; stipules persistent, subulate-lanceolate, 5–10 mm long; petiolar gland below first pair of leaflets saucer-shaped, reddish-brown, 0.5–1.5 mm in diameter. *Inflorescences*: bracted racemes of 1–6 flowers clustered in leaf axils; pedicels 1–2 cm long. *Flowers*: sepals minutely pubescent, lanceolate to lanceolate-acuminate, 6–14 mm long; petals 5, showy, bright yellow, 1–2 cm long, lowest petal larger, upper 4 with reddish-purple spot at base; 10 stamens, largest 8 mm long; pistil silky, arching away from the stamens; hypanthium scarcely evident. *Fruit*: legume, minutely pubescent, linear, flattened, straight or slightly curved, 4–6 cm long, 5–6 mm wide, mostly 9- to 15-seeded, dehiscing along two sutures, valves twisting spirally as they dry; seeds brownish-black, flattened, rectangular to rhomboidal, 3.5–4.5 mm long, with longitudinal rows of minute punctae. n=8.

This species is highly variable in size and pubescence. Several varieties have been described.

Synonyms: *Cassia fasciculata* Michx., *Chamaecrista fasciculata* (Michx.) Greene
Other Common Names: partridgepea, sensitive pea, locustweed

Showy partridgepea is infrequent to often abundant on disturbed prairies, bluffs, riverbanks and bottoms, and upland woods. It is most common along roadsides in infertile, sandy soils. It flowers from June to October. Domestic livestock consume showy partridgepea. A cathartic substance is present in the leaves and seeds. The substance is effective either in fresh plant material or in dry hay. Consumption of large quantities will cause stress and infrequently death. Deer eat it without being poisoned. Showy partridgepea forms excellent cover for upland wildlife. Numerous birds eat the seeds. Quail particularly seek its cover and seeds. It is an important honey plant. Nectar is not available in the flower but is supplied by the petiolar glands.

Leaves of showy partridgepea are sensitive to touch. Leaves quickly fold when handled.

Showy partridgepea is commonly grown as an ornamental. Seed is readily available, and germination is improved by scarification and stratification. It grows in association with rhizobial bacteria, and is sometimes used to improve soil fertility.

Cassia nictitans L., sensitive partridgepea, also is an annual. It has small petals (8 mm long or less) and only 5 stamens. It grows in the southeastern Great Plains.

19

Figure 11   *Cassia marilandica*

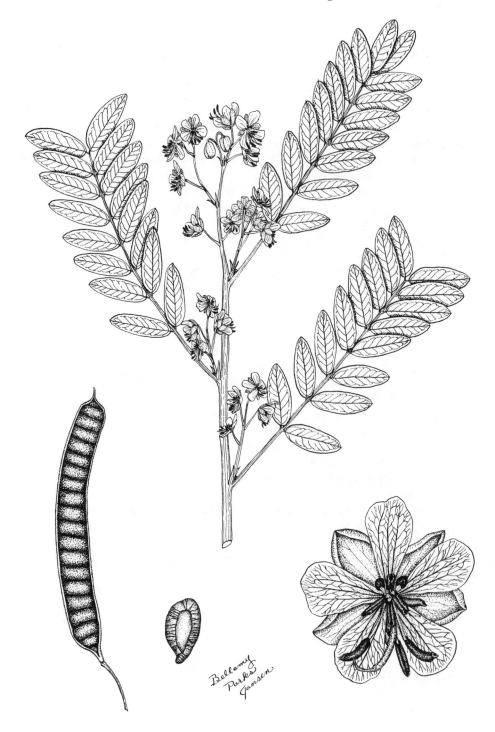

2. *Cassia marilandica* L. Wild senna (Figure 11)

[*marilandica*: of or from Maryland.]

Life Span: Perennial. *Origin*: native. *Height*: (0.5)1–2 m. *Stems*: herbaceous, smooth, from a woody caudex or horizontal rhizomes. *Leaves*: alternate, even-pinnately compound, 1–3 dm long; leaflets 8–24, elliptic or oblong, (1.5)2–8 cm long, (0.5)1.5–3 cm wide, glabrous, acute, apex mucronate, base slightly oblique; petiolar gland ovoid, near base of petiole; stipules caducous, linear-lanceolate, 7–10 mm long, 1 mm wide. *Inflorescences*: racemes, several-flowered, upper axillary and terminal. *Flowers*: sepals oblong to ovoid, 5–8 mm long, to 4 mm wide; petals 5, yellow, becoming cream-colored with brown veins when dry, obovate to obtriangular, 8–11(15) mm long, apex rounded or slightly emarginate; stamens 10, dark red, upper 3 erect and sterile, middle 4 fertile and projecting forward, lower 3 fertile and longer than the others; pistil silky, upcurved; ovary appressed-pubescent; hypanthium short. *Fruit*: legume, dark brown to black at maturity, glabrous (rarely sparsely pubescent), linear, flattened, straight or slightly curved, (4)6.5–10 cm long, 8–12 mm wide, tipped by persistent curved style, septate between the 10–25 seeds; seeds slightly flattened, oblong or obovate, 4.5–5.5 mm long, twice as long as wide, central portion dull, gray, rest of seed coat lustrous and darker gray. n=14.

Synonyms: *Cassia medsgeri* Shafer, *Ditremexa marilandica* (L.) Britt. & Rose
Other Common Name: Maryland senna

Wild senna is infrequent to common on roadsides, rocky banks, hillsides, and stream banks. It is most frequent in sandy, moist soils. It flowers from July through September. Domestic livestock occasionally eat the plant, and large quantities may cause poisoning.

Seeds are valuable food for birds. The petiolar glands furnish nectar for honey, and the flowers must be pollinated by insects. Seeds have a low rate of germination without scarification. Wild senna is not a desirable landscape plant, because it spreads rapidly by rhizomes. The cathartic value of leaves and pods, gathered after ripening, equals that of East Indian sennas commonly imported for the pharmaceutical trade.

Cassia roemeriana Scheele, twoleaved senna, has leaves with only 2 leaflets. It grows in the extreme southern Great Plains. Two other species, *C. obtusifolia* L. and *C. occidentalis* L., are rarely collected in the Great Plains.

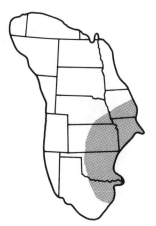

2. CERCIS L.

[*kerkis* (Gk.): ancient name of a tree, possibly the judas tree, *Cercis siliquastrum* L.]

Trees or shrubs with simple cordate or broadly obovate leaves with entire margins; calyx irregular, 5-lobed; corolla irregular, imperfectly papilionaceous, banner the smallest petal, keel petals the largest; stamens 10; hypanthium short, hemispheric, bearing the perianth and stamens on its margin; legume narrowly oblong, thin, flat, strongly margined along upper surface, several-seeded.

Seven species have been described in North America. Others are found in Eurasia. One species is common in the southeastern Great Plains.

Figure 12 *Cercis canadensis*

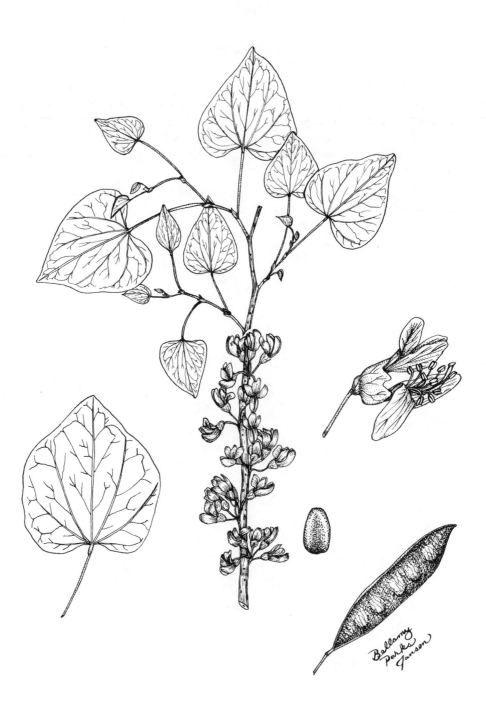

Cercis canadensis L.   Eastern redbud (Figure 12)

[*canadensis*: of or from Canada.]

Life Span: perennial. *Origin*: native. *Height and Form*: small tree to 8(12) m with a flat or rounded crown. *Twigs*: glabrous, often zig-zag. *Trunks*: to 3 dm in diameter. *Leaves*: deciduous, alternate, simple, palmately veined, usually with 5 principal veins, broadly cordate to broadly ovate, 8–14 cm long, 5–12 cm wide, tip acute or short-acuminate, margin entire, base cordate or truncate, dark green and glabrous above, paler with axillary tufts and a few hairs on the veins beneath; petiole 4–10 cm long, glabrous, enlarged at the summit and the base; stipules minute, caducous. *Inflorescences*: umbellike clusters of 4–8 flowers from lateral buds on old wood. *Flowers*: imperfectly papilionaceous, pink or rose (rarely white), 1.1–1.3 cm long; calyx tube irregularly campanulate, basally oblique, 2.5 mm long, 3–3.5 mm wide, 5-lobed; lobes broadly triangular to rounded, ciliate at the tip; banner 7–7.5 mm long, 3–3.5 mm wide; wings 7–7.8 mm long, strongly reflexed, enclose the banner in bud; keel petals distinct, 8–9 mm long; stamens 10; pedicels 8–17 mm long, slender, red, glabrous. *Fruit*: clustered legumes, remaining on the tree into early winter, pendulous, oblong, 5–10 cm long, 1.2–1.5 cm wide, slightly stipitate, flat, brown, glabrous, pointed at both ends, with a small ridge on both sides of upper suture, usually 5- to 12-seeded; seeds oval, 4–5 mm long, 4–4.5 mm wide, 2 mm thick, light to dark brown, semilustrous. 2n=14.

Other Common Names: judas tree, redbud

Eastern redbud is infrequent to common along the edge of woods or in open woods. It is also found on stream banks and hill-sides. It is grown widely beyond its native distribution. It flowers from March to May, before the leaves appear. Deer and cattle will browse young trees. Birds and squirrels eat the seeds.

Eastern redbud trees are readily available at nurseries. It is a popular landscape plant.

25

3. GLEDITSIA L.

[*Gleditsia*: named for J. G. Gleditsch (1714–1786), professor of botany at the University of Berlin.]

Trees or tall shrubs, usually armed with simple or branched thorns; leaves deciduous, alternate, evenly pinnate or bipinnate, leaflets crenulate; stipules minute, caducous; racemes spikelike, lax or dense; flowers unisexual or rarely perfect, small, greenish-yellow; calyx lobes 3–5, almost alike, inserted at the summit of the obconic hypanthium; petals 3–5; stamens 3–10, inserted at the base of the perianth; legumes large, oval to elongate, flat, scarcely dehiscent, 1- to many-seeded.

Twelve species have been described in North and South America, Africa, and Asia. One is common in the Great Plains.

Figure 13   *Gleditsia triacanthos*

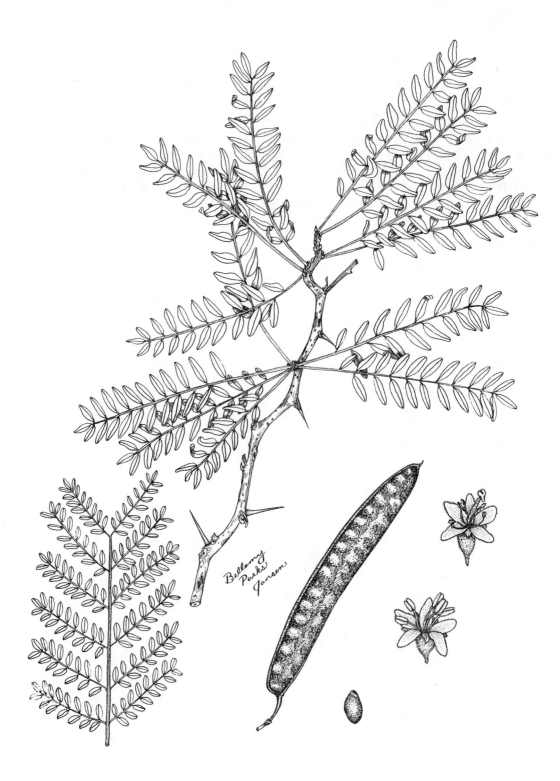

Gleditsia triacanthos L.

[*tri* (Lat.): three; + *akanthikos* (Gk.): of thorns, referring to the sometimes branched thorns.]

Life Span: perennial. *Origin*: native. *Height and Form*: tree to 15(25) m with a broad, open crown. *Twigs*: glabrous, reddish-brown to grayish-brown, zig-zag; thorns simple or branched, stout, shiny, reddish-brown, terete, flattened at the base, 4–10(20) cm long. *Trunks*: 6–9 dm in diameter. *Leaves*: deciduous, alternate, even-pinnately and bipinnately compound on the same tree; pinnate leaves 3- to 6-fascicled, from old wood, 1.5–3 dm long, with 18–32 leaflets; leaflets lanceolate to ovate, obscurely crenulate or entire, tip obtuse, often mucronate, base acute, glabrous above, pubescent on midrib beneath, sessile or with a 1 mm petiolule; bipinnate leaves from new growth, 2–4 cm long, with 4–16 pinnae and 10–24 leaflets per pinna; leaflets usually elliptic-oblong or lanceolate to narrowly ovate, 1–2.5 cm long, 4–8 mm wide; petioles 3–5 cm long, abruptly enlarged at the base, grooved above; stipules obsolete. *Inflorescences*: axillary racemes, usually solitary; staminate racemes 4–10 cm long, dense, many-flowered; pistillate racemes with fewer flowers. *Flowers*: staminate flowers in 3's, center flower opening first, greenish; calyx tube campanulate, (3)4(5) lobes; lobes linear, 2 mm long; petals 5, obovate, 2–2.2 mm long; stamens (3)4(10), distinct; pistillate flowers pedicellate; sepals 3(5), slightly smaller than the petals; petals 3(5); hypanthium short-campanulate. *Fruit*: pendant legumes, curved, twisted, 1–3(4.5) dm long, flat, 2.5–3 cm wide, base abruptly narrowed, tip acuminate or acute, walls thin, velvety pubescent when young and becoming glabrous, with 6–27 seeds; seeds oval, 9–12 mm long, 5–8 mm wide, hard, dull, brown, smooth, with subsurface fractures. 2n=28. Some plants do not have thorns. This horticultural selection is seldom found in the wild. The thornless type is forma *inermis* Schneid.

Other Common Name: common honeylocust

Honeylocust grows naturally in rich bottomlands, rocky hillsides, fencerows, and pastures. It is commonly planted as an ornamental and shade tree. It flowers from mid-May through June. Cattle and deer browse the young plants. Deer frequently strip and eat the soft bark of young trees in winter. Squirrels and many kinds of birds eat the seeds. It is also a source of honey.

Plants are readily available from commercial nurseries. In addition to the thornless trees, selections have been made for crown shape and foliage color. Honeylocust produces good firewood. It is also used in windbreaks to control erosion. The distribution, as shown on the map, has greatly increased in a northwesterly direction because of extensive planting.

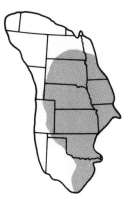

29

4. GYMNOCLADUS LAM.

[*gumnos* (Gk.): naked; + *klados* (Gk.): branch, referring to the absence of small branchlets.]

Tall, unarmed trees, without small twigs; bark rough, deeply fissured; leaves extremely large, compound; flowers in terminal panicles or racemes, perfect or partly perfect, greenish-white, regular; sepals 3–5, nearly like the petals, inserted in a single series at the summit of the tubular hypanthium; petals 3–5, oblong; stamens 10, distinct, alternating long and short; ovary sessile; legumes broadly oblong, hard, woody, thick, flat, containing a few large seeds, separated by pulp, tardily dehiscent, remaining on the tree until spring.

Only 2 species have been described. One grows in eastern Asia. The other is native to the eastern Great Plains.

Figure 14   *Gymnocladus dioica*

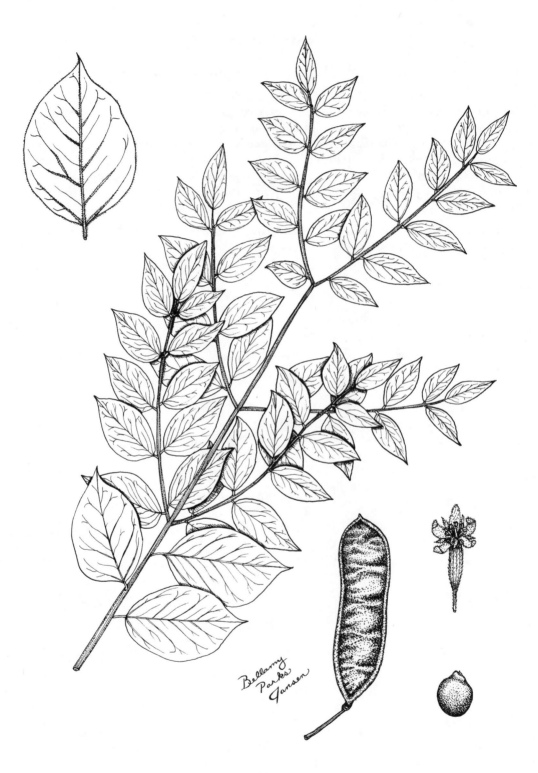

Gymnocladus dioica (L.) K. Koch Kentucky coffeetree (Figure 14)

[*di* (Gk.): two; + *oikos* (Gk.): house, referring to some of the plants being dioecious.]

Life Span: perennial. *Origin*: native. *Height and Form*: tree to 25(30) m with narrow shape and rounded crown, sometimes may form small colonies from root suckers. *Twigs*: stout, rigid, light brown, unarmed. *Trunk*: to 9 dm in diameter. *Leaves*: alternate, deciduous, bipinnately compound, 3–9 dm long, 3–6 dm wide; pinnae 6–14, 1–4 dm long, or a few pinnae replaced by a single large leaflet, with 10–14 leaflets per pinna; leaflets ovate, 4–7 cm long, 2–4 cm wide, apex acute to acuminate, base cuneate or obtuse and uneven, margin entire, upper surface pubescent near margin and on midrib, sparsely pubescent beneath; petiole 5–6 cm long; petiolules 2–3 mm long; stipules small, caducous. *Inflorescences*: lax, terminal panicles or racemes, staminate inflorescences 7–11 cm long; pistillate inflorescences often 2–3 dm long, with stout pedicels about 2–7 cm long. *Flowers*: imperfect or partly perfect; sepals 5, subequal; petals 3–5, obovate, greenish-white to pinkish-white, 4–5 mm long, with white pubescence outside; stamens 10, distinct, alternating long and short; hypanthium tubular-obconic, 6–10 mm long; 5 sepals, subequal. *Fruit*: stalked legumes, oblong, straight or slightly curved, heavy-walled, 5–15(25) cm long, 3–5(6) cm wide, purplish-brown to reddish-brown, 1- to 8-seeded, tardily dehiscent; seeds nearly circular, 1.5–2.1 cm long, 1.6–1.7 cm wide, hard, smooth, dark olive brown. 2n=28.

Other Common Names: nantita (Omaha-Ponca), tohuts (Pawnee), wahnahna (Lakota)

Kentucky coffeetree is scattered to locally common on rich alluvial soils of bottomlands. It occasionally grows on rocky hillsides. It was commonly planted on farmsteads. It flowers in May and June. Cattle, horses, sheep, and humans have been poisoned by Kentucky coffeetree leaves and pods. An alkaloid is probably responsible for the poisoning. Symptoms include gastrointestinal irritation and nervous manifestations that have been described as being narcotic in nature. Death is infrequent but can come within one day of ingestion. Sprouts growing from mowed areas are particularly a problem. Human poisoning has occurred following ingestion of the pulp between the seeds.

Indians on the Great Plains used the plant for a variety of purposes. Members of several tribes used a mixture of the pulverized bark and water for enemas. Seeds were roasted, ground, and used to make "coffee". Snuffed pulverized bark caused sneezing which was thought to relieve the pain of headaches. Lakota used the root to make a poor quality dye. Winnebago used the seeds for "counters" in gambling.

Kentucky coffeetree plants are available from nurseries. It produces durable, valuable wood. It has been used for erosion control in gullies, but other trees are better for this purpose.

5. HOFFMANSEGGIA CAV.

[*Hoffmanseggia*: named for J. C. von Hoffmannsegg (1766–1849), German botanist.]

Unarmed herbaceous or suffruticose plants; leaves odd-bipinnately compound, with 3 to several pinnae; leaflets mostly small, smooth, glandular-dotted; stipules persistent; racemes terminal or axillary, open; flowers perfect, nearly regular; calyx tube short, 5-lobed, valvate; petals 5, yellow or orange-red, conspicuous; stamens 10, distinct; legumes often glandular, straight or falcate, with few to several seeds.

A genus of about 30 species occurring mostly in dry areas of subtropical North and South America and South Africa. Several have extended into the southern United States from Mexico. Two grow in the Great Plains, and one is common in the southwestern portion.

Figure 15   *Hoffmanseggia glauca*

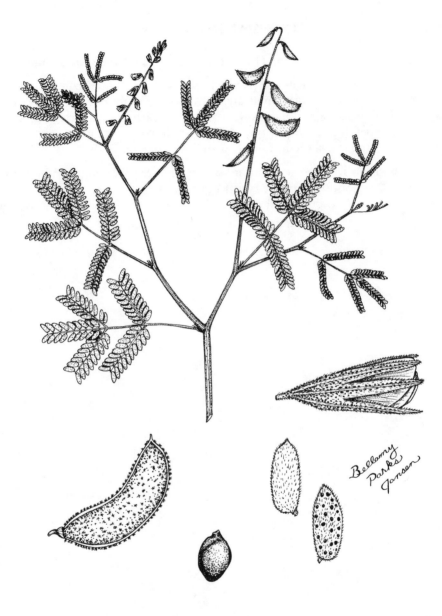

Hoffmanseggia glauca (Ort.) Eifert

Indian rushpea (Figure 15)

[*glaukos* (Gk.): silvery, in reference to the pubescence.]

Life Span: perennial. *Origin*: native. *Height*: 5–40 cm. *Stems*: herbaceous, branched or simple, spreading to erect, stipitate-glandular, pubescent to glabrate, from a caudex and fleshy tubers. *Leaves*: mostly basal or partly cauline, odd-bipinnately compound, with 3–15 pinnae and 12–22 leaflets per pinna; leaflets oblong to elliptic, 2–6 mm long, 2–3 mm wide, sessile, minutely pubescent to glabrate; petioles equal to or exceeding the rachis; stipules persistent, ovate-deltoid, ciliate. *Inflorescences*: terminal racemes of 4–17 flowers, 1–2 dm long, pubescent, stipitate-glandular. *Flowers*: calyx tube glandular, pubescent, 5-lobed; lobes 5.5–7 mm long; petals orangish-red, 1–1.3 cm long, with glandular claws; stamens 10, not exceeding the petals; pedicels glandular, pubescent, 2–5 mm long. *Fruit*: legumes flat, 2–4.5 cm long, persistent, indehiscent, shiny; seeds flat, green to brown.

Synonyms: *Hoffmanseggia densiflora* Benth. *ex* A. Gray, *H. galcaria* Cav., *Larrea densiflora* (Benth.) Britt.
Other Common Names: hog potato, pignut, rat sweetpotato

Indian rushpea is scattered to locally common in sandy or rocky prairie, abandoned fields, and roadsides. It is most common in alkaline soils. It flowers from May to September. It is palatable to livestock. There is no record of wildlife use.

Seed is not available from commercial sources. It is a valuable soil binder, but it becomes a serious weed. Its tuberous roots are edible and were commonly roasted and eaten by native Americans.

Hoffmanseggia drepanocarpa A. Gray, sicklepod rushpea, is similar, except that its stems and inflorescences are not stipitate-glandular. It also grows in the southwestern Great Plains.

FAMILY III. FABACEAE LINDL.

Trees, shrubs, twining vines, or herbs (annual or perennial); stems usually unarmed; leaves alternate, mostly palmately 3-foliate or pinnately compound (even or odd), rarely simple, never bipinnately compound, some with tendrils; commonly with stipules and stipels; inflorescences axillary or terminal, usually racemes; flowers variously colored, typically perfect and zygomorphic; calyx gamosepalous, typically 5-lobed, lobes usually unequal; petals usually 5, imbricate, papilionaceous; banner uppermost and outermost; wings lateral; keel petals lowermost, inside; stamens usually 10, commonly diadelphous or monadelphous (rarely distinct); ovary sessile or stipitate; fruit usually a legumes, straight, curved, or winged, dehiscent or seldom indehiscent; fruit sometimes a loment, separating into 1-seeded joints at maturity; endosperm absent.

Family containing up to 400 genera and 10,000 species. It is most common in the tropics, but it is well represented in temperate regions. Thirty-five genera and over 150 species have been recorded in the Great Plains, not including common garden plants. Seven genera not discussed here are *Clitoria, Galactia, Hedysarum, Indigofera, Pueraria, Sesbania,* and *Sphaerophysa.* Most of these are rather scattered and not common in the Great Plains.

Key to the Genera

A. Leaves simple, plants herbaceous
  B. Annuals; flowers yellow; stamens monadelphous.................8. *Crotalaria*
  B. Perennials; flowers purple to white; stamens diadelphous ....4. *Astragalus*
A. Leaves compound, plants herbaceous or woody
  C. Trees or shrubs
    D. Flowers not papilionaceous, petal 1; purple to blue; leaflets gland-dotted
    ....1. *Amorpha*
    D. Flowers papilionaceous
      E. Leaves even-pinnately compound; flowers yellow ....6. *Caragana*
      E. Leaves odd-pinnately compound; flowers white or rose to purple, sometimes partly marked with yellow
        F. Shrubs; flowers rose to purple with some yellow; legume less than 4 mm long ....9. *Dalea*
        F. Trees; flowers white; legume 5–10 cm long.............21. *Robinia*
  C. Herbaceous annuals and perennials
    G. Leaves even-pinnately compound, with tendrils
      H. Wing petals coherent with the keel petals; style filiform, bearded with a tuft or ring of hairs at the apex ....28. *Vicia*
      H. Wing petals free or nearly so; style flattened, bearded down the inner face ....12. *Lathyrus*
    G. Leaves odd-pinnately compound, palmate, or 3-foliate

I. Leaflets more than 3 on well developed leaves

J. Inflorescence an umbel or capitate cluster
    K. Inflorescence many-flowered (8 or more), pink to white; fruit a loment . . . .
    . . . . .7. *Coronilla*
    K. Inflorescence few-flowered (8 or fewer), orangish-red or white; fruit a legume. . .
    . . . . .14. *Lotus*
J. Inflorescence a terminal, axillary, or lateral raceme or spike
    L. Leaves palmately compound
        M. Leaflets 5 to 11; foliage not gland-dotted; legumes 1-seeded. . . . . . .15. *Lupinus*
        M. Leaflets 5 (3 on upper leaves); foliage gland-dotted; legumes several-seeded
        . . . . .20. *Psoralea*
    L. Leaves pinnately compound
        N. Leaves glandular-dotted or glandular-punctate
            O. Flowers in terminal spikes; legumes without hooked prickles; stamens mon-
            adelphous . . . . .9. *Dalea*
            O. Flowers in axillary racemes; legumes with hooked prickles; stamens di-
            adelphous . . . . .11. *Glycyrrhiza*
        N. Leaves not glandular
            P. Plants twining; calyx lobes unequal, the lowest lobe surpassing the others
            . . . . .3. *Apios*
            P. Plants not twining; calyx lobes all similar in length
                Q. Inflorescences terminal on leafy stems
                    R. Calyx lobes shorter than the tube; stamens distinct; legumes con-
                    stricted between the seeds, terete .22. *Sophora*
                    R. Calyx lobes equal to or longer than the tube; stamens monadelphous;
                    legumes linear, flattened . . . . .25. *Tephrosia*
                Q. Inflorescences axillary or borne on scapes
                    S. Legumes short (8 mm long or less), flat, one-seeded, with short, blunt
                    spines on the margins . . . . .18. *Onobrychis*
                  S. Legumes ovoid or subglobose to linear, not spiny
                        T. Keel rounded to acute. . . . . .4. *Astragalus*
                      T. Keel abruptly narrowed to a prolonged beaklike appendage
                      . . . . .19. *Oxytropis*
I. Leaves 3-foliate
    U. Leaves serrulate (apparently entire in some *Trifolium*)
        V. Terminal leaflet sessile or nearly so . . . . .27. *Trifolium*
        V. Terminal leaflet distinctly stalked
            W. Inflorescences of globose heads, short spikes, or short racemes
                X. Flowers blue to white, yellow, or cream; legumes elongate, flat, curved or
                coiled; corolla deciduous after withering . . . . .16. *Medicago*
                X. Flowers yellow; inflorescences capitate or of short racemes; legumes
                short, 1- or 2-seeded
                    Y. Corolla persistent after withering . . . . .27. *Trifolium*
                    Y. Corolla deciduous after withering. . . . . .16. *Medicago*
            W. Inflorescences of elongate racemes; flowers yellow or white; legumes sub-
            globose, 1- or 2-seeded . . . . .17. *Melilotus*

U. Leaves entire
- Z. Flowers usually yellow (sometimes white to light purple), in open peducled racemes
  - AA. Foliage blackening on drying; legume inflated ....5. *Baptisia*
  - AA. Foliage not blackening on drying; legume flat, elongate......26. *Thermopsis*
- Z. Flowers white, cream, blue, purple, or pink, in globose heads, spikes, racemes, panicles, or solitary
  - BB. Flowers cream, usually with a purple spot on the banner ......13. *Lespedeza*
  - BB. Flowers white, blue, purple, or pink
    - CC. Plants with stems normally twining
      - DD. Inflorescence a raceme; chasmogamous and cleistogamous flowers in separate locations; keel petals straight ....2. *Amphicarpaea*
      - DD. Inflorescence a subcapitate raceme; flowers all chasmogamous; keel petals strongly curved.....................23. *Strophostyles*
    - CC. Plants with stems erect to procumbent, not twining
      - EE. Flowers in terminal racemes or spikes
        - FF. Flowers in dense cylindrical spikes ....9. *Dalea*
        - FF. Flowers in racemes
          - GG. Fruit a loment; terminal leaflet distinctly stalked
          ....10. *Desmodium*
          - GG. Fruit a legume; terminal leaflet sessile or with a stalk no longer than stalks of lateral leaflets...........5. *Baptisia*
      - EE. Flowers axillary, solitary or clustered
        - HH. Flowers solitary (rarely in 2's)
          - II. Flowers pink; fruit a glabrous legume..........14. *Lotus*
          - II. Flowers orangish-yellow to white; fruit a 2-segmented loment...................................24. *Stylosanthes*
        - HH. Flowers several to many in a cluster
          - JJ. Foliage glandular-punctate ....20. *Psoralea*
          - JJ. Foliage not glandular.....................13. *Lespedeza*

1. AMORPHA L.

[*amorphos* (Gk.): no form or deformed, in reference to the flowers having only one petal.]

Erect shrubs or suffrutescent shrubs, sometimes from rhizomes; leaves alternate, odd-pinnately compound; leaflets numerous, entire (occasionally crenate), commonly mucronate; inflorescences of dense, spike-like racemes, from terminal and upper axils; flowers pedicellate, small; calyx obconic or turbinate, usually persistent; calyx lobes 5, acuminate to obtuse, equal or upper lobes shortest; the single petal (banner) obovate, purple to blue (rarely white), wrapped around stamens and style, clawed; stamens 10, monadelphous at the base, otherwise distinct; ovary short; legumes often curved, longer than the calyx, gland-dotted, 1- to 2-seeded.

About 20 species in North America, most numerous in the southern United States. Only 3 species are commonly reported in the Great Plains.

A. Tall shrubs, usually 1.5–4 m tall; petioles 2–5 cm long; leaflets green, 2–4.5 cm long
....2. *A. fruticosa*
A. Low suffrutescent shrubs, rarely over 1 m tall; leaves nearly sessile; leaflets less than 2 cm long
  B. Plant gray-canescent throughout; racemes several to many in a cluster
....1. *A. canescens*
  B. Plant mostly glabrous, foliage green; racemes usually solitary, terminal
....3. *A. nana*

Figure 16   *Amorpha canescens*

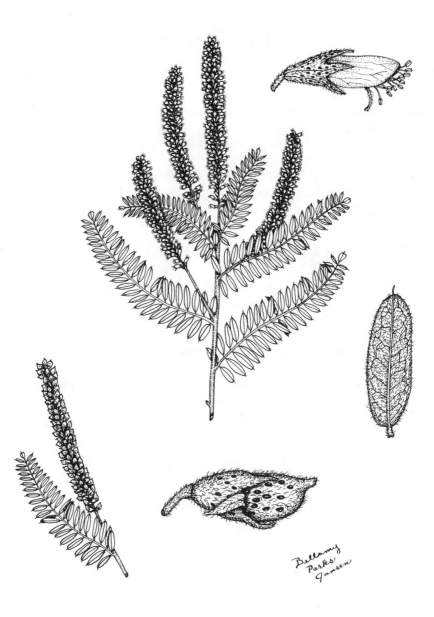

1. *Amorpha canescens* Pursh   Leadplant (Figure 16)

[*canescens* (Lat.): becoming gray, in reference to its covering of fine hairs.]

Life Span: perennial. *Origin*: native. *Height*: 3–10 dm (may be 20 dm in ungrazed and protected areas). *Stems*: suffrutescent, erect (sometimes ascending), often from rhizomes, new growth tomentose. *Leaves*: 3.5–10 cm long, odd-pinnately compound; leaflets 15–51, crowded to overlapping, elliptic to oblong, entire, (7)8–14(18) mm long, 3–6 mm wide, apex mucronate, lower surface canescent, light green, upper surface not as woolly, darker green; petioles tomentose, 1–2(5) mm long; stipules subulate, 1–3 mm long, caducous. *Inflorescences*: several terminal racemes, central raceme the longest and first to flower, 6–10 cm long; rachis tomentose. *Flowers*: bright purple (occasionally light blue to violet blue), calyx tube 3–5 mm long, turbinate with resinous glands, 5-lobed, 1–1.6 mm long, lanceolate; petal 1, broadly obovate with a slender claw, 4–5 mm long, 3 mm wide, incurved and enclosing stamens and pistil; stamens 10; anthers yellow to yellowish-orange, conspicuous; ovary densely pilose; pedicels tomentose, 0.5–1.6 mm long. *Fruit*: legume, curved, tomentose-canescent, 3–4.8 mm long, 1.6–2 mm wide, 1-seeded, calyx persistent; seeds with slight beak, elliptic, 2–2.8 mm long, 1–1.5 mm wide, orange-brown, smooth. 2n=20.

Synonym: *Amorpha brachycarpa* E. J. Palm.
Other Common Names: prairie shoestring, downy amorpha, false greasewood, wild tea, zitka tacan (Lakota), te huto hi (Omaha-Ponca)

Leadplant is infrequent to abundant on well-drained prairie and open woods. It flowers from June through early August.

It is excellent forage for livestock and wildlife. It is high in nutritive quality, and animals select it before eating most other species. Leadplant decreases with heavy grazing. It is rarely abundant on overgrazed rangeland and is, therefore, an important indicator of range condition.

In mowed and burned prairies, it has the appearance of a large herbaceous plant. Seed is commercially available. Germination is enhanced with scarification and stratification. It can also be propagated from softwood stem cuttings. It is easily transplanted and is becoming increasingly popular as a landscape plant.

Indians on the Great Plains dried the leaves for smoking in pipes and for tea. They also used it to treat neuralgia and rheumatism. They would cut stems into small pieces and attach one end of each piece to the skin over the affected area by first moistening it. Then the pieces of stem were lighted and allowed to burn down to the skin as a counterirritant.

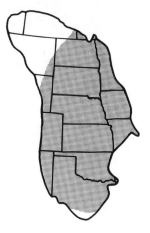

Figure 17   *Amorpha fruticosa*

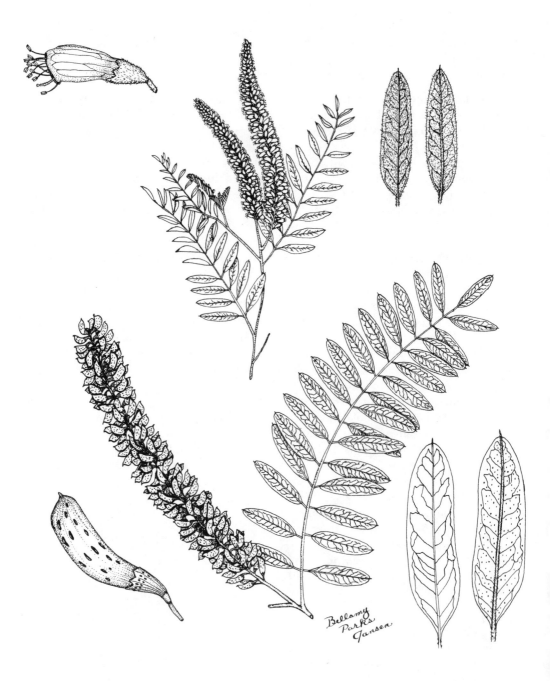

2. *Amorpha fruticosa* L. False indigo (Figure 17)

[*fruticosus* (Lat.): shrubby, for its general appearance.]

Life Span: perennial. ***Origin***: native. ***Height and Form***: shrub 1.5–3(5) m tall. ***Stems***: erect with a single stem or a few clustered trunks, sometimes from rhizomes, finely ribbed branching near the top; bark of young trunks smooth, brownish-gray, with prominent transverse lenticels; bark of old trunks slightly fissured. ***Leaves***: alternate, 12–22 cm long, odd-pinnately compound; leaflets elliptic to oblong, 2–4.5 cm long, 8–18 mm wide, entire, apex obtuse, generally mucronate (0.2–0.4 mm long), base acute or obtuse, upper surface glabrous (especially when mature) to appressed pubescent and dull dark-green, lower surface variously hairy (less when mature) and glandular-punctate; stipules linear to subulate, pubescent, 2–4 mm long, caducous. ***Inflorescences***: racemes, terminal, usually in 2's or 3's, densely flowered, 10–15 cm long. ***Flowers***: purple (rarely white to blue); calyx tube obconic, 2–4 mm long, glabrate, the 5 calyx lobes acute to broadly rounded, 0.5–1.2 mm long, upper lobes shortest, pubescent, sometimes glandular; petal 1, broadly ovate, 5–6 mm long, folded to enclose stamens and pistil, apex obtuse, claw indistinct; stamens 10, 6–8 mm long, anthers yellow to yellowish-orange. ***Fruit***: legume, 6–8 mm long, 1.5–3 mm wide, brown, curved, strongly glandular, 1-seeded; seeds tan to brown, slight beak, oblong-oval, 3–4.5 mm long, 1.2–1.5 mm wide, smooth, glossy. $2n=40$.

Synonyms: *Amorpha angustifolia* (Pursh) Boynt., *A. fragrans* Sweet, *A. fruticosa* var. *angustifolia* Pursh and var. *tennesseensis* (Shuttlew.) E. J. Palm.
Other Common Names: indigobush, indigobush amorpha, river locust, stream-bank amorpha, water string, kitsuhast (Pawnee), sunktawote (Lakota)

False indigo is infrequent to locally common on moist stream banks, in open woods, prairie gullies, rocky banks, and along lake and pond shores. It flowers as early as April in the south, but most flowering occurs in May and June. It is consumed by livestock, although it generally is not present in large enough quantities to be an important component of their diet. Wildlife utilize the foliage and seeds. It decreases with heavy grazing.

Commercial seed is seldom available. It can be started from softwood cuttings. False indigo was formerly planted for windbreak borders and in buffer strips to prevent erosion. These plantings were generally not successful. It is sometimes used for roadside plantings and commercial and residential landscaping.

Pawnee gathered the leafy branches, whenever possible, and spread them on the ground near the place of butchering. Fresh meat was placed on the branches to keep the meat clean. Lakota cut false indigo to feed to horses and made arrow shafts from the stems.

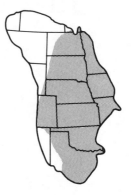

Figure 18   *Amorpha nana*

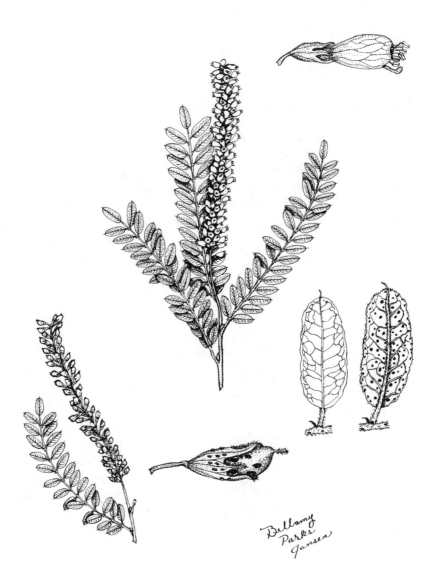

3. *Amorpha nana* Nutt.

Dwarf wildindigo (Figure 18)

[*nanos* (Gk.): dwarf or small, describing the plant's stature.]

Life Span: perennial. *Origin*: native. *Height and Form*: low shrub, 4–6(10) dm tall. *Stems*: suffrutescent, erect (occasionally ascending), usually 1 main stem with numerous branches, rhizomatous; bark gray, split longitudinally, slightly rough. *Leaves*: alternate, odd-pinnately compound, 3–8(10) cm long; leaflets 25–41, (2)7–11(15) mm long, 2–6 mm wide, crowded, elliptical to broadly oblong, margins shallowly crenate, often ciliate, apex obtuse, usually mucronate (1.5 mm long); upper surface glabrous and dark yellowish-green; lower surface lighter, glandular-punctate, sparingly pubescent; petiole 3–8(10) mm long, pubescent; stipule 1, subulate, 3–5 mm long, reddish-brown gland on side of petiole opposite the stipule; stiple minute. *Inflorescences*: raceme, solitary, terminal, 3–9 cm long, 1–2 cm wide, densely flowered. *Flowers*: calyx tube obconic, punctate-glandular, 2 mm long; calyx lobes 5, 1.2 mm long, acuminate, upper lobes shorter, ciliate; petal 1, purple with darker purple veins, obovate, 4.5 mm long, 4 mm wide, apex obtuse and usually emarginate, with a slender claw enveloping stamens and pistil; stamens 10, filaments united below; ovary glabrous. *Fruit*: legume, 4–6 mm long, 2–5 mm wide, reddish-brown, glabrous, pubescent at tip, glandular, 1-seeded, the calyx persistent; seed narrowly oval, 2.5 mm long, 1.5 mm wide, flattened, olive-brown.

Synonym: *Amorpha microphylla* Pursh
Other Common Names: dwarf leadplant, fragrant falseindigo

Dwarf wildindigo is scattered, but locally common, on dry prairies and rocky or sandy hillsides. It generally grows in the open and flowers from late May through July. It is often overlooked on prairie, because it may be browsed down to 15 cm. It decreases with heavy grazing. Forage quality is probably similar to that of leadplant (*Amorpha canescens*), although no specific data are available.

Generally, seed is not commercially available. It can be used for landscaping. Transplanting and propagation from seeds or softwood cuttings have been successful.

2. AMPHICARPAEA Ell. *ex* Nutt.

[*amphi* (Gk.): of both kinds; + *karpos* (Gk.): fruit, in reference to 2 kinds of fruit.]

Low perennial herbs, twining; leaves pinnately trifoliate, stipulate; leaflets subtended by small stiples; flowers of 2 kinds, those of upper branches chasmogamous, in axillary racemes, with purple to white petals, each pedicel subtended at base by a striate-veined bractlet; calyx slightly irregular; stamens diadelphous, style elongate, legume valves coiled after dehiscence; flowers at base of plant cleistogamous, borne on threadlike branches; petals rudimentary; stamens few.

Only 7 species have been described. Six species are found in east Asia. One species grows in the eastern Great Plains.

Figure 19   *Amphicarpaea bracteata*

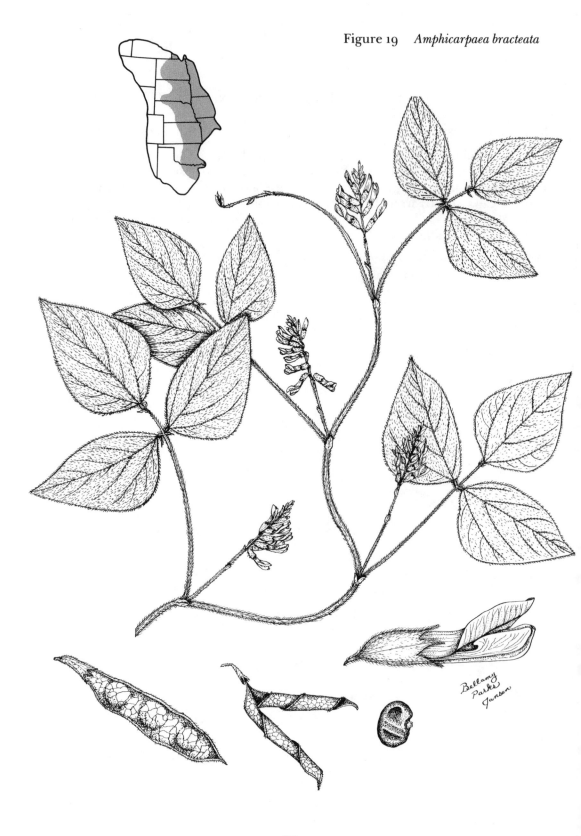

Bellamy
Parks
Jansen

Amphicarpaea bracteata (L.) Fern.   Hog peanut (Figure 19)

[*bractea* (Lat.): covered with thin metal plates, in botany applied to the thin, scalelike leaves subtending flowers.]

Life Span: perennial. *Origin*: native. *Height*: twining vine, 3–20 dm long. *Stems*: herbaceous, densely hairy, from taproots. *Leaves*: alternate, pinnately 3-foliate; leaflets ovate to broadly ovate (occasionally rhombic), 2–7(10) cm long, 1.5–7 cm wide, apex acute or acuminate, base of center leaflet obtuse, lateral leaflets oblique, both surfaces pubescent; petioles 2–7 cm long; terminal petiolule 5–40 mm long, lateral petiolules 1–2.5 mm long; stipules membranaceous, 3–8 mm long, persistent, pubescent. *Inflorescences*: racemes, 5- to 20-flowered, individual or pairs of flowers subtended by an obtriangular bract; pedicels 1–6 cm long. *Flowers*: of two types; chasmogamous flowers with corolla papilionaceous, pale purple to white (rarely red); calyx tube 5–7 mm long, persistent; calyx lobes 4, lanceolate to deltoid, 0.5–2 mm long; banner 1.2–1.4 cm long, margins recurved; wing and keel nearly straight; cleistogamous flowers produced near plant base or subterranean, lacking well-developed petals. *Fruit*: aerial legume straight or slightly curved, 1.5–4 cm long, 7 mm wide, usually 3(2–4)-seeded, valves pubescent, coiled after dehiscence; seeds reddish-brown, mottled with black, round to somewhat 3-sided, slightly flattened, 3–6 mm long; subterranean legumes fleshy, 6–12 mm in diameter, 1-seeded. 2n=20.

Hog peanut varies in size of leaflets and pubescence. Var. *comosa* (L.) Fern., coarse plants, and var. *bracteata*, slender plants, are sometimes recognized.

Synonyms: *Amphicarpa comosa* (L.) Nieuw. & Lunnell, *A. pitcheri* T. & G.

Other Common Names: American hogpeanut, hairy hogpeanut

Hog peanut is scattered in brushy ravines, woods, and thickets. It generally twines on herbs and shrubs, but it is occasionally found growing on bare, moist banks. It flowers from August through September. The vines are eaten by cattle, and the underground legumes are rooted up and consumed by hogs.

Seed is not commercially available. Hog peanut has no potential for landscaping.

Hog peanuts are important food for many types of wildlife. Seeds and subterranean legumes are consumed by song birds and upland game birds, as well as by deer. Seeds are considered to be important locally for wintering quail and pheasants. Many species of rodents depend on hog peanuts throughout the winter.

The subterranean legumes are edible after cooking. They are nutritious after ripening in September and October. They can be excavated throughout the winter. They were an important source of food for Indians in the upper Missouri Valley, as well as for early travelers and settlers. Indians searched for rodent caches of the legumes and would dig them out to be used for food. Up to 2 liters of legumes could be obtained from one wood rat cache. Indians usually replaced the cache with an equal amount of corn.

3. APIOS Fabr.

[*apios* (Gk.): a pear, in reference to the occasional shape of the tubers.]

Twining, milky-juiced perennial herbs from rhizomes with fleshy tubers; leaves alternate, odd-pinnately compound, with 5–7 stalked leaflets; flowers in axillary racemes, each flower subtended by a pair of linear, 1-nerved, caducous bractlets; calyx 5-lobed, with 4 short lobes, 1 slightly longer; banner round or obovate, short-clawed, auricled; wings curved, oblong or obovate, auricled, deflexed below the keel; keel strongly curved, blunt-tipped; stamens 10, diadelphous; style curved with the keel; legumes linear, several-seeded, splitting into two valves, coiled after dehiscence.

A genus consisting of 8 species, 6 of which occur in eastern Asia. Two species grow in North America, and 1 is found in the Great Plains.

Figure 20   *Apois americana*

[*americana*: of or from America.]

Life Span: perennial. *Origin*: native.
Height: climbing, twining vine, 1–3(5) m
long. *Stems*: herbaceous from rhizomes,
smooth, sparsely pubescent (generally not
visible without magnification); rhizomes
bearing 2 to many (up to 40) fleshy tubers,
oblong to globose, 1–2 cm thick. *Leaves*:
alternate, odd-pinnately compound; leaflets (3)5–7(9), lanceolate to ovate, 2–6(10)
cm long, 1–4 cm wide, acute or acuminate,
mucronate, glabrous to sparsely pubescent; petioles 1.5–8 cm long; stipules 4–7
mm long, subulate, caducous. *Inflorescences*: dense axillary racemes, 5- to many-
flowered; each node of the inflorescence
subtended by 2 minute bracts; pedicels 1–
4 mm long. *Flowers*: calyx tube 4–6 mm
long, glabrous to sparsely pubescent,
5-lobed; upper 4 lobes small, lower lobe
prominent; corolla papilionaceous; banner broader than long, usually retuse,
white outside, red to brown inside, 9–13
mm long, apically notched; wings
brownish-purple, down-curved; keel purple or green. *Fruit*: legumes linear,
straight or slightly curved, 5–10 cm long,
5–6 mm wide, few-seeded, valves coiling
after dehiscence, seeds dark brown,
plump, oblong, slightly flattened, surface
wrinkled, 4–5 mm long. 2n=40.

Synonyms: *Apios tuberosa* Moench, *Glycine
apios* L.
Other Common Names: American potato-
bean, wildbean, Indian potato, blo
(Lakota)

Groundnut is infrequent to locally common in moist soils of prairie ravines, pond
and stream banks, and thickets. The vines
are eaten by domestic livestock, especially
horses, and the tubers are eaten by hogs.

Seeds are eaten by upland game birds and
song birds. Tubers are eaten by mice, rabbits, and squirrels.

Seed is not commercially available. The
plant has no potential for landscaping, although it holds promise as a tangle vine
for erosion control.

The starchy tubers furnished food for
Indians and early settlers. Indians gathered tubers in the fall, often taking them
from rodent caches. A few tribes cultivated groundnut. Tubers were roasted or
boiled, tasting like potatoes. Seeds were
used as peas are used today. The Pilgrims
were forced to rely heavily on groundnuts
during their first few hard winters.

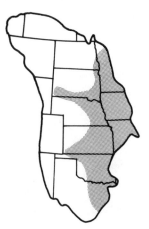

4. ASTRAGALUS L.

[*astragalos* (Gk.): ancient name for some leguminous plant.]

Annual or perennial; stems herbaceous, caulescent or not, simple or branching; from a rhizome, caudex, or taproot; leaves odd-pinnately compound (sometimes palmately 3-foliate and rarely reduced to one leaflet); leaflets several to numerous; petioles absent or present; stipules connate or distinct; axillary racemes loose or dense, elongated or contracted; flowers white, ochroleucous, yellow, pink, or purple; calyx tube campanulate to cylindric, often oblique, 5-lobed; lobes short, triangular or subulate; banner obovate-oblong to rotund, clawed, usually reflexed from the wings, usually exceeding the wings; wings clawed, apex usually notched or rounded; keel shorter than the wings, obtuse to acute, clawed; stamens 10, diadelphous; legumes short or long, sessile to conspicuously stipitate, turgid or flattened, thick-walled or thin-walled, mostly fleshy in texture, sometimes the lower margin or both margins depressed or intruded as a false partition, the fruit thus is bilocular or unilocular, dehiscent (rarely indehiscent), 1- to many-seeded, subtended by the remnants of the stamen tube and ruptured calyx.

A wide-spread genus containing about 1,500 species, which are mainly located in the northern temperate and arctic zones. It is a complex, highly variable, and difficult genus. Mature legumes are often necessary for recognition of the species. Some species, often called locoweeds, produce an alkaloid-like substance that is toxic to livestock (see discussion with *Astragalus mollissimus*), and some, the poisonvetches, accumulate toxic levels of selenium (see discussion with *Astragalus bisulcatus*). Of the 37 species reported for the Great Plains, 23 that are among the most important and abundant are included in the key.

A. Leaves primarily simple; some lower leaves may be compound
    B. Leaves spatulate; legumes narrowly oblong and solid-colored, mostly basal, subacaulescent . . . . .22. *A. spatulatus*
    B. Leaves linear-filiform; legumes inflated and mottled, caulescent. .5. *A. ceramicus*
A. Leaves all compound
    C. Leaves palmately 3-foliate
        D. Flowers pinkish-purple (rarely white); banner 7 mm long or less 21. *A. sericoleucus*
        D. Flowers white to yellow (tips of wings and keel rarely light purple); banner 1.1 cm long or more
            E. Petals dorsally villous; banner narrowed in the middle; calyx tube 5–7 mm long . . . . .12. *A. hyalinus*
            E. Petals glabrous; banner tapering from tip to base; calyx tube 7–15 mm long . . . . .10. *A. gilviflorus*

C. Leaves 5- to many-foliate
  F. Leaflets spine-tipped . . . . .13. *A. kentrophyta*
  F. Leaflets not spine-tipped
G. Pubescence of the leaves mostly dolabriform
  H. Plant obviously caulescent
    I. Legumes glabrous; inflorescences shorter than the subtending leaves
      . . . . .4. *A. canadensis*
    I. Legumes strigulose; inflorescences longer than the subtending leaves
      . . . . .1. *A. adsurgens*
  H. Plants tufted to short-caulescent
    J. Flowers of 2 kinds (chasmogamous and cleistogamous); banner 1.4 cm long or
      less; calyx tube 5 mm long or less . . . . .14. *A. lotiflorus*
    J. Flowers of 1 kind (chasmogamous); banner 1.5 cm long or more; calyx tube 6
      mm long or more . . . . .15. *A. missouriensis*
G. Pubescence on leaves basifixed
  K. Stems and leaves villous with hairs 1–2 mm long from minutely bulbous bases
    . . . . .8. *A. drummondii*
  K. Stems and leaves variously pubescent with hairs shorter and without bulbous
    bases
    L. Stipules distinct
      M. Banner less than 1 cm long; legumes pubescent or glabrous . . . .
        . . . . .17. *A. nuttallianus*
      M. Banner 1.5 cm long or more; legumes glabrous to puberulent
        N. Plant caulescent; inflorescences shorter than subtending leaves; foliage
          strigulose. . . . . .7. *A. crassicarpus*
        N. Plant subcaulescent to short-caulescent; inflorescences about equaling
          the subtending leaves; foliage villous-tomentose . . . . .16. *A. mollissimus*
    L. Stipules connate at least at lower nodes
      O. Banner 1 cm long or less
        P. Fruit laterally compressed; calyx tube usually with black hairs
          . . . . .23. *A. tenellus*
        P. Fruit terete or angular, not compressed; calyx tube without black hairs
          Q. Leaflets linear, 2 mm wide or less; fruit less than 1 cm long
            . . . . .11. *A. gracilis*
          Q. Leaflets oblanceolate, 2 mm wide or more; fruit more than 1 cm long
            . . . . .9. *A. flexuosus*
      O. Banner more than 1 cm long
        R. Leaflets sessile, or nearly so
          S. Fruits bilocular and inflated; banner less than 1.5 cm long . . .6. *A. cicer*
          S. Fruits unilocular and slightly compressed; banner more than 1.5 cm
            long . . . . .18. *A. pectinatus*
        R. Leaflets with petiolules
          T. Legumes sessile or with stipe 1.5 mm long or less, usually bilocular
            U. Racemes 4 cm long or less; pedicels less than 2 mm long; legumes
              1 cm long or less . . . . .2. *A. agrestis*
            U. Racemes 4 cm long or more; pedicels 2 mm long or more; le-
              gumes 1–2 cm long . . . . .19. *A. plattensis*

T. Legumes with stipe longer than 1.5 mm, unilocular
    V. Legume strongly 2-grooved on the ventral face; calyx lobes often
       red ....3. *A. bisulcatus*
    V. Legume not grooved ventrally; calyx lobes green 20. *A. racemosus*

Figure 21   *Astragalus adsurgens*

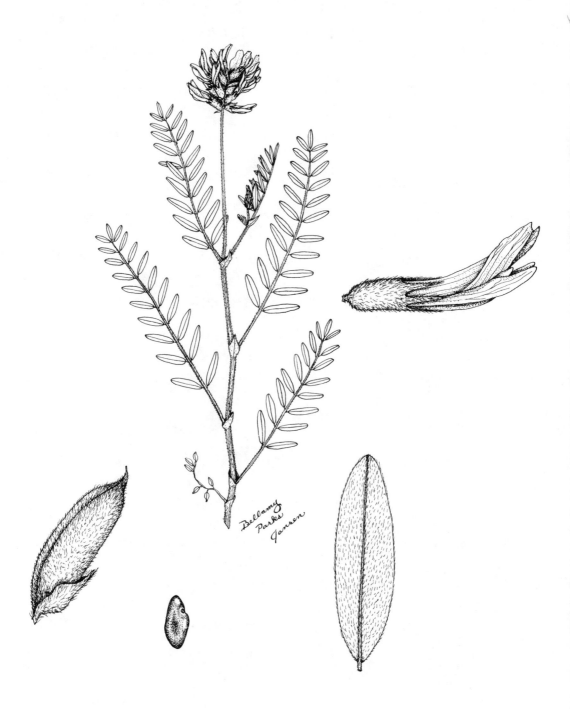

1. *Astragalus adsurgens* Pall.　　　　　　Prairie milkvetch (Figure 21)

[*adsurgens* (Lat.): ascending, in reference to most of the stems.]

Life Span: perennial. *Origin*: native.
Height: 1–3(4) dm. *Stems*: herbaceous,
erect to decumbent, cespitose, clustered,
simple, from woody taproots, strigulose,
with dolabriform hairs. *Leaves*: alternate,
odd-pinnately compound, 4–15 cm long;
leaflets 9–25, elliptic-lanceolate to oblong-
obovate, 8–25 mm long, 3–9 mm wide,
apex acute, glabrate or strigose above;
lower stipules connate for at least 1/3 of
their length, 5–15 mm long. *Inflorescences*: axillary racemes, subcapitate to
densely spicate, 2–10 cm long, 12- to 50-
flowered; peduncles 4–14 cm long.
Flowers: calyx tube cylindric, 5–8 mm
long, strigulose with white and black hairs,
5-lobed; lobes 1.5–5 mm long; corolla
blue, lavender, purple, or white; banner
scarcely recurved, 1.2–2 cm long, 4–8 mm
wide; wings 9–17 mm long, claws 4–8 mm
long; keel 8–15 mm long, claws 4–8 mm
long; stamens 10, diadelphous; pedicels
less than 1 mm long. *Fruit*: legumes erect
or ascending, largely or partially concealed by the persistent calyx, substipitate
to sessile, ellipsoid to ellipsoid-lanceolate,
triquetrous to subterete, 7–12 mm long,
2.5–4 mm wide, dorsally sulcate, valves
thickly papery, bilocular or nearly so,
closely strigulose with basifixed hairs, few-
to many-seeded; seeds smooth, brown to
nearly black, 2–2.8 mm long. n=16.

Var. *robustior* Hook. is the variety common to the Great Plains. It is highly variable in flower color and amount of pubescence.

Synonyms: *Astragalus chandonetti* Greene,
A. striatus Nutt.
Other Common Name: standing milkvetch

Prairie milkvetch is locally common on
prairies, sand hills, roadsides, and rocky
areas. It flowers from May to September.
It is not palatable to livestock and is seldom by eaten by wildlife.

Commercial seed is seldom available. It
is infrequently used for landscaping.

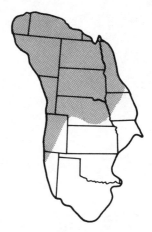

Figure 22   *Astragalus agrestis*

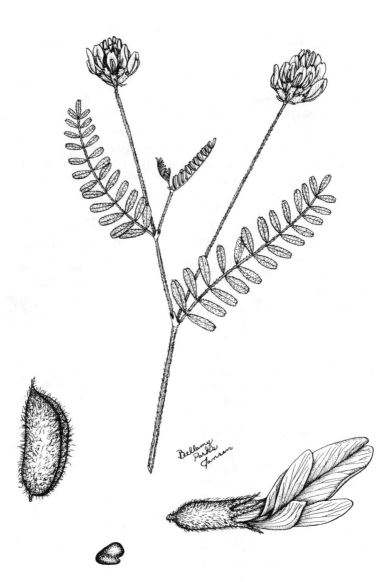

2. *Astragalus agrestis* Dougl. *ex* G. Don

Field milkvetch (Figure 22)

[*agrestis* (Lat.): pertaining to land, in reference to the habitat.]

Life Span: perennial. ***Origin***: native. ***Height***: 1–3(4) dm. ***Stems***: herbaceous, decumbent to ascending, slender, glabrous or sparsely strigose with hairs basifixed, mostly single from slender and elongate rhizomes. ***Leaves***: alternate, odd-pinnately compound, 3–12 cm long, subsessile or petiolate; leaflets 9–27, linear-oblong or elliptic-oblong to lanceolate, 4–20 mm long, apex obtuse to retuse, sparsely pilose on both sides; stipules on lower part of the stem connate, linear to ovate, 2–9 mm long. ***Inflorescences***: axillary racemes, dense, ovoid to subcapitate, 2–4 cm long, scarcely elongating, 5- to 20-flowered; peduncles longer or shorter than the leaves. ***Flowers***: calyx tube short-cylindric, 4–8 mm long, more or less black-hirsute, 5-lobed; lobes 3–5 mm long, subulate; corolla pinkish-purple, blue, lavender (sometimes white, ochroleucous, or yellow); banner slightly reflexed, oblanceolate to elliptic, notched, 1.6–2.3 cm long, wings 1.4–1.8 cm long, claws 6–8 mm long; keel 1–1.4 mm long, claws 6–9 mm long; stamens 10, diadelphous; pedicels 1–1.5 mm long. ***Fruit***: clustered legumes, nearly sessile, obliquely ovoid, 7–10 mm long, 3–5 mm wide, densely hirsute with white hairs, deeply sulcate on lower margin, obcompressed or somewhat triquetrous, bilocular; seeds brown to black, smooth, 1.5–2 mm long. n=8.

Synonyms: *Astragalus dasyglottis* Fisch. *ex* DC., *A. goniatus* Nutt., *A. hypoglottis* Hook.

Field milkvetch is infrequent to locally common in moist prairies and meadows, along margins of lakes and streams, on open wooded hillsides, in thickets, and along roadsides. It flowers from May to August. Domestic livestock sometimes consume it, but its palatability is not high.

Commercial seed is rarely available. It occasionally is used for landscaping.

Figure 23 *Astragalus bisulcatus*

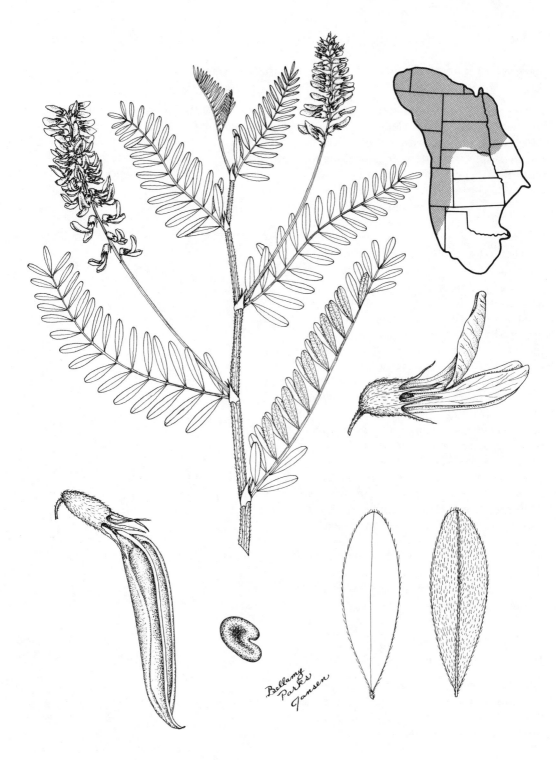

Bellamy
Parks
Jansen

3. *Astragalus bisulcatus* (Hook.) A. Gray Twogrooved poisonvetch (Figure 23)

[*bi* (Lat.): two; + *sulcus* (Lat.): furrow, referring to the two furrows, or grooves, in the legume.]

Life Span: perennial. *Origin*: native. *Height*: 2–8 dm. *Stems*: few to many from a caudex and taproot, herbaceous, ascending to erect (sometimes decumbent), clump-forming, strigulose or glabrate with basifixed hairs. *Leaves*: alternate, odd-pinnately compound, 4–13 cm long, petioled below on the stem to sessile above; leaflets 13–29(35), ovate to elliptic below, elliptic to linear above, acute or obtuse and sometimes mucronate, usually glabrate above, strigulose beneath; lower stipules connate, upper stipules distinct, scarious, 3–10 mm long. *Inflorescences*: axillary racemes, dense to lax, ovoid to cylindric, many-flowered, flowers reflexed; peduncles 3–12 cm long. *Flowers*: calyx tube gibbous, campanulate to subcylindric, strigulose, 3.5–6 mm long, 5-lobed; lobes often red, (1.5)2–4(5) mm long; corolla purple, white with purple maculate keel, or entirely white (to lavender or pink); banner obovate to oblanceolate, 1–1.7 cm long, keel 6–14 mm long, wings 8–16 mm long; stamens 10, diadelphous; pedicels 1–3 mm long. *Fruit:* legumes pendulous, stipitate or exserted-stipitate (stipe 1.5–5 mm long), oblong or shortly ellipsoid, 7–20 mm long, 2–5 mm wide, obcompressed, unilocular, strongly 2-grooved on the ventral face, many-seeded, valves thick-papery, smooth or transversely rugose, strigulose to glabrate; seeds yellow or brown to black, smooth, 3–3.5 mm long. n=12.

Synonyms: *Astragalus diholcos* Tidest., *A. haydenianus* A. Gray, *Diholcos bisulcatus* (Hook.) Rydb.

Other Common Names: twogrooved milkvetch, twogrooved locoweed.

Twogrooved poisonvetch is locally common in prairies, badlands, stream valleys, and roadsides. It only grows on seleniferous soils. It flowers from May to August. Reports on its palatability vary. It is sometimes preferentially grazed in Montana, but most reports from the Great Plains indicate that it is not readily grazed.

Twogrooved poisonvetch accumulates selenium. Animals consuming single, massive amounts of herbage may exhibit blindness, wandering, excitement, and depression. These symptoms are followed by coma, respiratory failure, and death.

Commercial seed is not available. It has been introduced into horticulture. It is not valuable for soil conservation.

Astragalus alpinus L., alpine milkvetch, is similar, but the stems arise from buried points. It grows only in the Black Hills of South Dakota in the Great Plains. *Astragalus bodini* Sheld., Bodin milkvetch, is also similar. But, its stipules are all connate, and its banner is smaller. It grows in the western Great Plains, where it is not common.

Figure 24 *Astragalus canadensis*

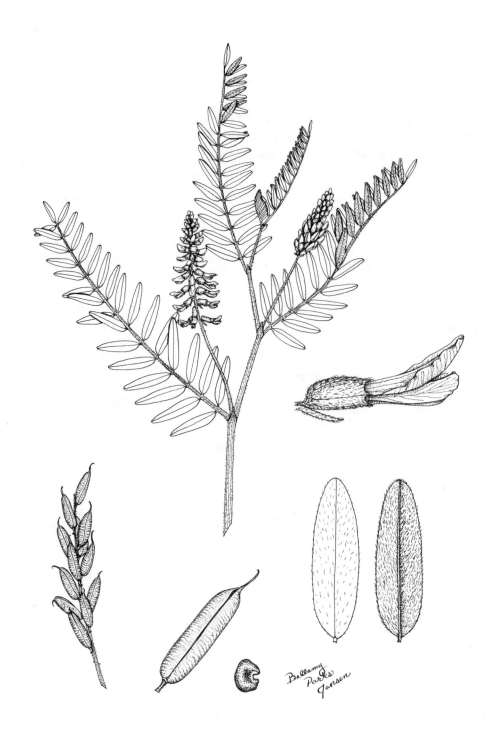

4. *Astragalus canadensis* L.   Canada milkvetch (Figure 24)

[*canadensis*: of or from Canada.]

Life Span: perennial. **Origin**: native. **Height**:1.5–15 dm. **Stems**: herbaceous, erect, robust, few, often branched, solid or hollow, from somewhat oblique rhizomes, glabrous to thinly strigose with dolabriform pubescence. **Leaves**: alternate, odd-pinnately compound, 5–35 cm long, sessile to short-petioled; leaflets 13–35, oblong to elliptic to narrowly lanceolate, 1–4 cm long, 5–15 mm wide, on short petiolules, apex obtuse, base obtuse or acute, glabrous or rarely strigose above, more or less strigose beneath, the pubescence dolabriform with one arm longer than the other; stipules connate, membranous, lanceolate to deltoid, 3–12(18) mm long. **Inflorescences**: axillary racemes, dense, 3–20 cm long, many-flowered; flowers spreading or declined, imbricate; peduncle 4–10 cm long, shorter than the subtending leaves. **Flowers**: calyx tube 4–7 mm long, campanulate to broadly cylindric, strigose or glabrate, 5-lobed; lobes acuminate 1–2(5) mm long; corolla greenish-white or ochroleucous, sometimes tinged with purple, often maculate on the keel; banner 1.1–1.7 cm long; wings 1–1.5 cm long, clawed; keel 9–13 mm long, clawed; stamens 10, diadelphous. **Fruit**: legumes numerous, ascending or erect, crowded, ovoid to oblong, 1–1.8 cm long, 5–6 mm wide, subterete in cross-section, bilocular, filiform-beaked, tardily dehiscent, both surfaces convex or dorsal surface sulcate, valves coriaceous, rugose, glabrous (rarely puberulent); seeds brownish-yellow, smooth, obliquely cordate to reniform, 2–2.5 mm long. 2n=16.

Synonyms: *Astragalus carolinianus* L., *A. halei* Rydb.

Other Common Name: pejuta ska hu (Lakota)

Canada milkvetch is locally common in moist prairies, open woodlands, roadsides, thickets, and stream banks. It flowers from May to August. It is palatable to livestock. Deer also eat the herbage. Horses have been reported to selectively graze the fruit.

Canada milkvetch has been introduced into horticulture, and seeds are commercially available. It is valuable for erosion control.

Indians on the Great Plains pulverized and chewed the roots for back and chest pains. Tea was made from the roots to control coughing.

Astragalus miser Dougl. *ex* Hook. var. *hylophilus* (Rydb.) Barneby, woodland milkvetch, is similar in appearance. It has unilocular fruits and is infrequent in the Black Hills of South Dakota.

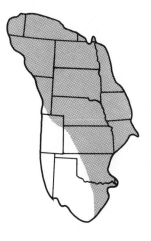

Figure 25 *Astragalus ceramicus*

5. *Astragalus ceramicus* Sheld.

Birdegg milkvetch (Figure 25)

[*keramos* (Gk.): a vessel, referring to the shape and coloration of the legume.]

Life Span: perennial. *Origin*: native. *Height*:0.5–4 dm. *Stems*: herbaceous, decumbent to ascending, sometimes diffusely foliose, flexuous, simple or few-branched from a caudex or from extensive rhizomes, strigulose pubescence with basifixed hairs (rarely dolabriform). *Leaves*: alternate, mostly simple and reduced to phyllodes, otherwise the lower ones odd-pinnately compound, 4–15 cm long; leaflets where developed, elliptically oblong to involute-filiform, 5–30 mm long; phyllodes linear-filiform, usually curved or somewhat hooked at the apex; petioles of compound leaves 2–9 cm long; lower stipules connate. *Inflorescences*: axillary racemes, loose, 2–7(15) cm long, 2- to 7-flowered, ascending, spreading, deflexed; peduncles 1–7 cm long. *Flowers*: calyx tube campanulate, 2–4 mm long, white or black strigose, 5-lobed; lobes 1–2(3) mm long; corolla white to pink (rarely purple), usually with the keel maculate; banner recurved, 6–9 mm long, clawed; wings 6–8 mm long, clawed; keel 6–8 mm long, clawed; stamens 10, diadelphous. *Fruit:* legumes pendulous, stipitate (stipe 1–3 mm long) or sessile, unilocular, inflated and bladder-like, ellipsoid to broadly ovoid, subterete, 1.5–5 cm long, 1–2.5 cm wide; valves papery, glabrous, mottled with red or purple; seeds smooth, brown, 2–3 mm long.

The most abundant variety in the Great Plains is var. *filifolius* (A. Gray) Herm. It is most common in the Nebraska Sandhills.

Synonyms: *Astragalus longifolius* (Pursh) Rydb., *Phaca longifolius* (Pursh) Nutt.
Other Common Names: painted milkvetch, tasusu canhologan (Lakota)

Birdegg milkvetch is found almost exclusively in sand. It is rare to locally common in prairies, hills, dunes, blowouts, and roadsides. It flowers from June to July. Livestock occasionally eat it, but it is unimportant because it is small and usually scattered. Birds and rodents consume the seeds.

Seed of birdegg milkvetch is not available from commercial sources. It has a potential horticultural value from the novelty of the inflated legumes. It has no value for erosion control.

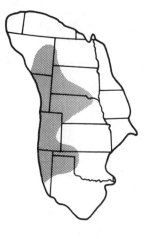

71

Figure 26 *Astragalus cicer*

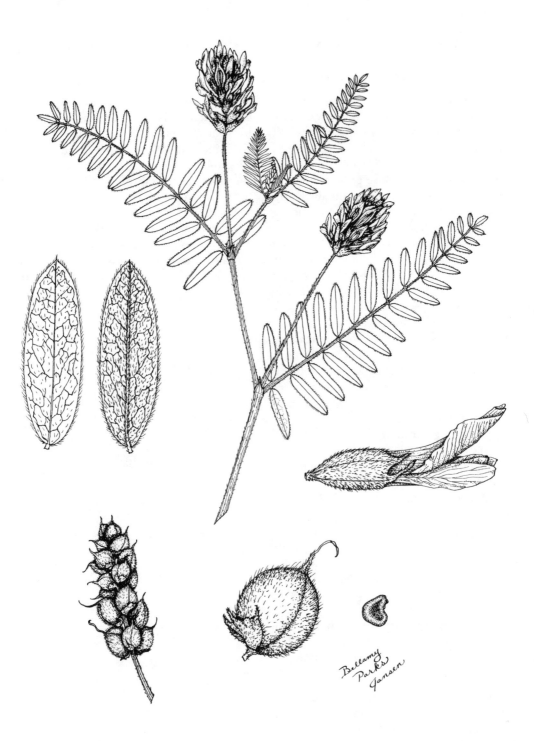

6. *Astragalus cicer* L.

Cicer milkvetch (Figure 26)

[*cicer* (Lat.): a chick-pea in reference to its papilionaceous flowers.]

Life Span: perennial. *Origin*: introduced (from Europe). *Height*: 3–10 dm (to 30 dm long). *Stems*: several to many from short and stout rhizomes; herbaceous, decumbent to ascending, tangle-forming, hollow, strigulose to glabrate with basifixed hairs. *Leaves*: alternate, odd-pinnately compound, 5–20 cm long, sessile; leaflets 17–27(35), elliptic to oblong, retuse to acute, usually obtuse sometimes mucronate, 5–30 mm long, 2–10 mm wide, often glabrate above, sparsely pubescent beneath; lower stipules connate. *Inflorescences*: axillary racemes, shorter than or slightly surpassing the subtending leaves, compactly ovoid to short-spicate, 10–20 cm long, with 10–60 ascending flowers. *Flowers*: calyx tube cylindric to campanulate, 4–5 mm long, 5-lobed; lobes sometimes irregular, 1–2 mm long; corolla pale yellow to white or ochroleucous; banner oblanceolate, reflexed, 1.2–1.5 cm long, wings rounded or emarginate, 1.1–1.4 cm long, keel incurved, 1–1.3 cm long; stamens 10, diadelphous; pedicels in flower 0.3–0.8 mm long, in fruit 0.5–1.5 mm long. *Fruit:* legumes clustered, spreading to ascending, sessile to substipitate, ovoid to ellipsoid, inflated, bilocular, subterete, 1–1.5 cm long, ventrally sulcate, valves thickly papery, coriaceous, pustulate-hirsute; seeds bright yellow to pale green, ovate to reniform, shiny.

Synonym: *Astragalus galegiformis* in herbaria pro parte

Cicer milkvetch was introduced to North America from Sweden in 1926. It is planted under dryland and irrigated conditions in the western Great Plains. It has rarely escaped to roadsides and fields. It flowers in June and July. Cicer milkvetch is eaten by all classes of livestock either as hay or pasture. It is also grazed by deer, elk, and pronghorn. No cases of bloat have been recorded. Cicer milkvetch contains no harmful alkaloids, nor does it accumulate selenium in toxic amounts.

A few cultivars are available. Seed requires scarification and should be innoculated with rhizobial bacteria before planting. Cicer milkvetch is best adapted to cool, moist sites with moderately coarse-textured soil. It withstands slightly acidic to moderately alkaline conditions and grows well under irrigation. Cicer milkvetch is valuable for critical area plantings, soil conservation, and wildlife habitat. It is occasionally grown as an ornamental novelty.

Figure 27   *Astragalus crassicarpus*

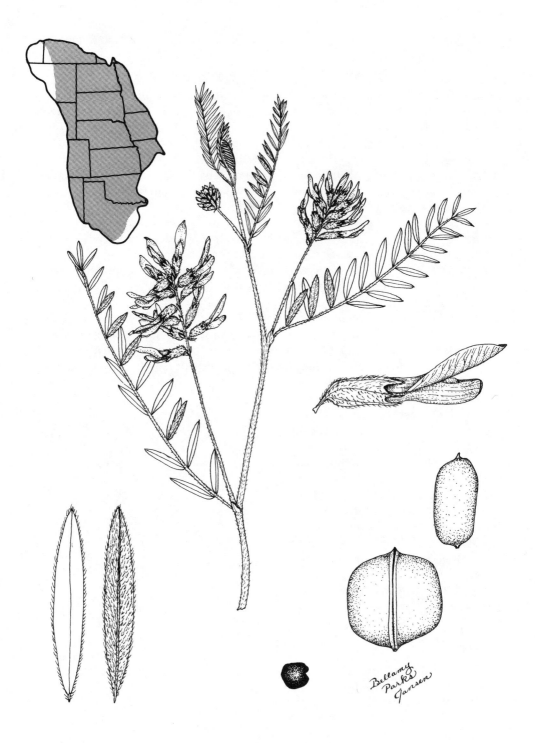

7. *Astragalus crassicarpus* Nutt.

Groundplum milkvetch (Figure 27)

[*crassus* (Lat.): thick; + *karpos* (Gk.): fruit, in reference to its plum-like fruit.]

Life Span: perennial. *Origin*: native. *Height*: 1–6 dm. *Stems*: several, clustered from a strong caudex and taproot, herbaceous, ascending to prostrate, glabrate to strigulose with basifixed hairs. *Leaves*: alternate, odd-pinnately compound, 2–10(14) cm long, mostly subsessile; leaflets 11–33, oblanceolate to elliptic, 3–17 mm long, 2–6 mm wide, apex obtuse to acute, often mucronate, cuneate at the base, appressed pilose above and beneath or glabrate above; stipules somewhat membranous, lanceolate, distinct, 3–9 mm long. *Inflorescences*: axillary racemes, shorter than or equal to subtending leaves, 5- to 25-flowered with the flowers ascending or spreading. *Flowers*: calyx tube subcylindric, 6–9 mm long, strigulose to tomentose with black and white hairs, 5-lobed; lobes acuminate or long-acuminate, 1–4 mm long; corolla purple to blue to pinkish-white, or greenish-white to ochroleucous; banner moderately reflexed, 1.5–2.5 cm long, notched, exceeding the wings and keel; wings 1.4–1.8 cm long; keel clawed; stamens 10, diadelphous. *Fruit:* legume subglobose, 1.5–4 cm long, 2 cm in diameter, with a filiform beak, valves thick-walled, glabrous, bilocular, initially succulent, green or reddish-purple on the upperside, many-seeded; seeds black, smooth or pitted, 2–4 mm long. 2n=22.

Three varieties occur in the Great Plains. The simplest way of separating them is by color of the flowers. Var. *crassicarpus* has purple or blue petals, and var. *paysonii* (Kelso) Barneby has white petals and a purple-tipped keel. Var. *trichocalyx* (Nutt.) Barneby has greenish-white to ochroleucous petals.

Synonyms: *Geoprumnon crassicarpum* (Nutt.) Rydb., *G. mexicanum* (A. DC.) Rydb., *G. succulentum* (Richards.) Rydb., *G. trichocalyx* (Nutt.) Rydb., *Astragalus caryocarpus* Ker, *A. mexicanus* A. DC., *A. succulentus* Richards., *A. trichocalyx* Nutt.

Other Common Names: buffalo bean, buffalo pea, ground plum, bi'jikiwi'bugesan (Chippewa), pe ta wote (Lakota)

Groundplum milkvetch is scattered to locally common on rocky or sandy prairie hillsides and uplands. It flowers from March to May. It is not rated as highly palatable, but it does decrease with continued heavy grazing. The legumes are cached by rodents.

Commercial seed is not available. Groundplum milkvetch is occasionally grown as a novelty. It is not valuable for erosion control.

Pioneers ate raw or cooked pods, and made them into pickles. Lakota ate the fruits and said that they were good medicine for horses. Chippewas used the fruits in compounding medicines for convulsions and hemorrhages from wounds. For an unknown reason, Pawnee placed its fruits in the water in which they placed their seed corn before planting.

Astragalus distortus T. & G., Ozark milkvetch, also has decumbent stems with purple corollas. Its stems are glabrous and the banner 1.5 cm long or less. It occurs in the southeastern Great Plains.

Figure 28 *Astragalus drummondii*

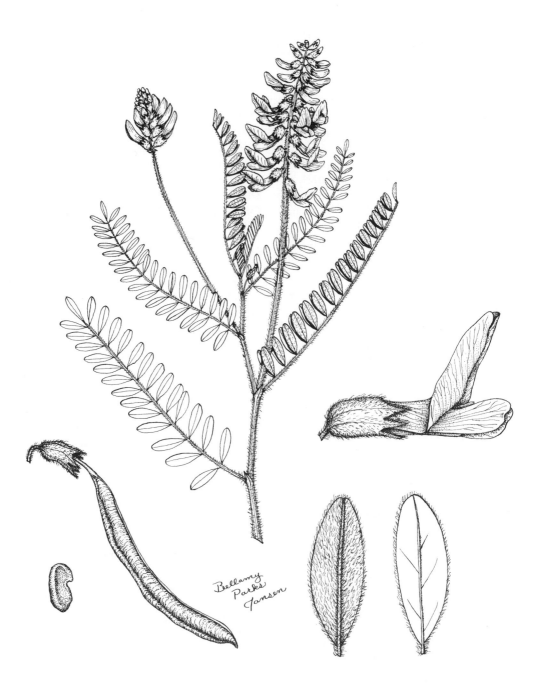

8. *Astragalus drummondii* Dougl. *ex* Hook. Drummond milkvetch (Figure 28)

[*drummondii*: named for James Drummond, English botanist, who explored North America in the second quarter of the 19th century.]

Life Span: perennial. *Origin*: native. *Height*:3–5(7) dm. *Stems*: herbaceous, erect or ascending, robust, hollow, usually clustered, simple from short rhizomes, villous, with spreading hairs 1–2 mm long from bulbous bases. *Leaves*: alternate, 5–12 cm long, all but lowest subsessile, leaflets 12–33, sometimes not paired, elliptic-oblong to oblanceolate, 5–25 mm long, obtuse or truncate, glabrous to glabrate above, glabrate to villous beneath with basifixed hairs; lowermost stipules connate or distinct, the remainder distinct, 3–12 mm long. *Inflorescences*: axillary racemes, extended beyond the subtending leaves, 12- to 36-flowered, flowers spreading to declined; peduncles 5–13 cm long. *Flowers*: calyx tube subcylindric to campanulate, 4–8 mm long, white or black (or mixed) pubescence, 5-lobed; lobes 2–4 mm long; corolla white to pale yellow with a purple-tipped or maculate keel; banner 1.5–2.5 cm long; wings 1.5–2 cm long, claws 6–9 mm long; keel 1.2–1.5 cm long, claws 6–8 mm long; stamens 10, diadelphous. *Fruit:* legumes pendulous, usually exserted with a stipe 3–12 mm long, sometimes included, oblong to oblanceolate, straight or slightly curved, 1–3 cm long, 3–5 mm wide, obcordate or sub-triquetrous in cross-section, bilocular or nearly so, unilocular at the tip, dorsally sulcate; valves thickly papery, glabrous; seeds brown, reniform, shiny to dull, 2–3 mm long.

Synonym: *Tium drummondii* (Dougl.) Rydb.

Drummond milkvetch is frequent to locally common in prairies, roadsides, foothills, and stream terraces. It grows in a wide range of soil textures. It flowers from May to July. Its palatability is rated as fair or lower because of its coarseness. Deer and pronghorn occasionally graze the foliage.

Seed is not available from commercial sources. Drummond milkvetch has little potential for landscaping or erosion control.

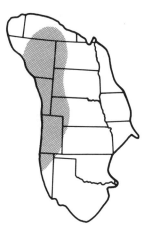

Figure 29   *Astragalus flexuosus*

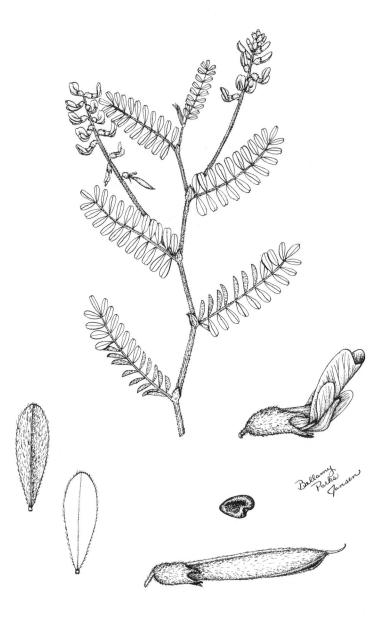

9. *Astragalus flexuosus* (Hook.) G. Don — Flexile milkvetch (Figure 29)

[*flexuosus* (Lat.): winding, bending, in reference to the pliant stems.]

Life Span: perennial. ***Origin***: native. ***Height***:(1)3–6 dm. ***Stems***: solitary or few from a caudex or stout rhizome, herbaceous, decumbent to spreading ascending, thinly canescent, the pubescence straight or incurved, basifixed. ***Leaves***: alternate, odd-pinnately compound, 3–8 cm long, petiolate below to subsessile above; leaflets (11)15–27, oblanceolate or oblong to linear, 5–15 mm long, apex truncate to retuse, glabrous above (rarely pubescent), strigose beneath; stipules connate below, distinct above, 1.5–7 mm long. ***Inflorescences***: axillary racemes, 2–3 cm long at anthesis, much longer at maturity, 10- to 26-flowered; the flowers spreading, then declining; peduncles 2–7 cm long. ***Flowers***: calyx tube campanulate, 2.5–4 mm long, canescent, 5-lobed; lobes triangular, 0.5–1.5(2) mm long; corolla pale purple to white; banner 7–10 mm long, recurved; wings 7–9 mm long; keel usually acute, recurved, 5–6 mm long; stamens 10, diadelphous. ***Fruit***: legumes widely spreading or reflexed, sessile or on a stipe to 2 mm long (much shorter than the persistent calyx), oblanceolate or ellipsoid, straight or slightly curved, turgid or inflated, not sulcate, unilocular, valves papery, pubescent, reticulate to cross-rugose; seeds pale brown, smooth, 2–2.5 mm long. n=11.

Synonyms: *Astragalus stictocarpus* (Rydb.) Tidest., *Pisophaca elongatea* (Hook.) Rydb., *P. flexuosa* (Dougl.) Rydb.
Other Common Name: pliant milkvetch

Flexile milkvetch is infrequent to common on prairies, roadsides, and bluffs. It is most abundant on sandy clay to shale soils. It flowers from May to August. Flexile milkvetch is grazed by all classes of livestock, but it is generally rated as rather low in palatability. Deer and pronghorn also graze the foliage.

Flexile milkvetch seed is not commercially available. It has little potential for use in landscaping or erosion control.

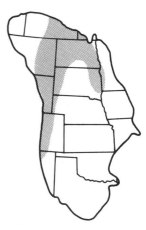

79

Figure 30   *Astragalus gilviflorus*

10. *Astragalus gilviflorus* Sheld.

Threeleaf milkvetch (Figure 30)

[*gilvus* (Lat.): pale yellow; + *floris* (Lat.): flower, referring to the flower color.]

Life Span: perennial. ***Origin***: native. ***Height***: 2–14 cm. ***Stems***: several from a cespitose caudex and a taproot, very short, herbaceous, the plant mounded or tufted, white-strigose with dolabriform hairs. ***Leaves***: 3–15 cm long, 3-foliate; leaflets usually narrowly oblanceolate to rhombic-obovate or obovate-cuneate, acute to acuminate (rarely obtuse), strigose; petioles elongated, strigose; stipules oblong-ovate to ovate, occasionally transversely corrugated, 5–14 mm long. ***Inflorescences***: axillary racemes, 1- to 3-flowered, capitate, shorter than subtending leaves; peduncles obsolete. ***Flowers***: calyx tube cylindric, 7–15 mm long, 5-lobed; lobes 1–5 mm long, long-acuminate; corolla white (rarely ochroleucous), becoming yellow on drying, keel tip purple (keel rarely all blue); banner 1.5–2.9 cm long, tapering from tip to base; wings 1.2–2.5 cm long; keel 1–2.2 cm long, claws 8–16 mm long; stamens 10, diadelphous; pedicels 0–1.5 mm long; calyx subtended by a linear-lanceolate bract. ***Fruit:*** legumes erect, ovoid to ellipsoid, 5–12 mm long, 3–6 mm wide, strigulose, beaked, usually 1-seeded; seeds yellowish-brown to black, sometimes purple, broadly reniform, smooth, 1.5–2 mm wide. n=12.

Synonyms: *Astragalus triphyllus* Pursh, *Orophaca caespitosa* (Nutt.) Britt.
Other Common Name: plains orophaca

Threeleaf milkvetch is scattered to locally common on barren slopes, hilltops, and flats. It grows on a variety of soils. It flowers from April to May. It is grazed by all classes of livestock, as well as by pronghorn.

Threeleaf milkvetch has been introduced into horticulture, but seed is not available from commercial sources. It has little value for erosion control.

Astragalus barrii Barneby, Barr's milkvetch, also has palmately 3-foliate leaves. Its calyx tube is 5 mm long or less. It is relatively rare and is found primarily in western South Dakota, western Nebraska, and eastern Wyoming.

81

Figure 31 *Astragalus gracilis*

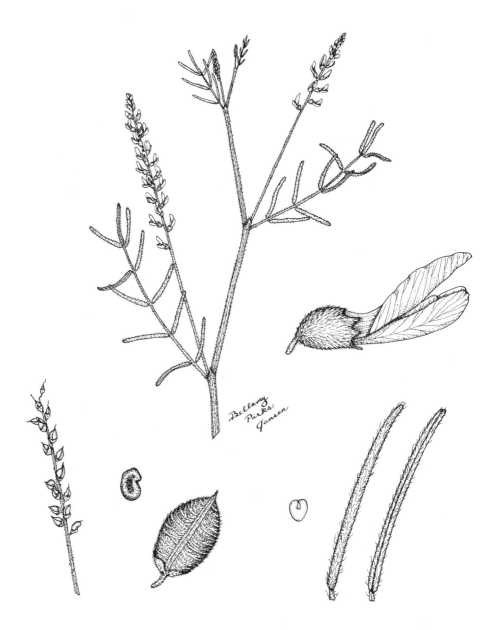

11. *Astragalus gracilis* Nutt.

Slender milkvetch (Figure 31)

[*gracilis* (Lat.): slender, referring to the stems and leaves.]

Life Span: perennial. ***Origin***: native. ***Height***:1.5–4(8) dm. ***Stems***: herbaceous, erect to nearly prostrate, slender, much branched from the base (rarely simple), strigulose (rarely glabrous), from short rhizomes. ***Leaves***: alternate, odd-pinnately compound, 2–9 cm long; leaflets 7–21, linear to narrowly oblong, 4–25 mm long, 1–2(4) mm wide, often involute, apex retuse to truncate or obtuse, glabrate above, strigulose beneath with basifixed hairs; petioles 0–3 cm long; stipules of upper leaves connate, papery, 1–4 mm long. ***Inflorescences***: axillary or terminal racemes, longer than the subtending leaves, slender, lax, 5- to 50-flowered; peduncles naked, 4–22 cm long. ***Flowers***: calyx tube campanulate, 1.5–2.5 mm long, 5-lobed; lobes slightly irregular, angular or short acuminate, 0.5–1 mm long; corolla light purple or lavender to dark purple (rarely white) with a purple-tipped keel; banner moderately to strongly reflexed, 6–9 mm long; wings 5–8 mm long; keel incurved, 4–6 mm long; stamens 10, diadelphous; pedicels in flower 0.5–1 mm long, in fruit to 2 mm long. ***Fruit***: legumes spreading and eventually drooping, ellipsoid to broadly ovoid, 4–9 mm long (including a short beak), 3–5 mm wide, unilocular, inflated; valves strigose; coriaceous, cross-rugose; 5- to 9-seeded; seeds brown or green, smooth, minutely pitted, 2.2–3.5 mm long.

Synonyms: *Microphacos gracilis* (Nutt.) Rydb., *M. parviflora* (Pursh) Rydb.
Other Common Name: pejuta skuya

Slender milkvetch is locally common on dry, rocky, or sandy prairie hillsides, uplands, stream valleys, and bluffs. It flowers from May to September. Domestic livestock, deer, and pronghorn occasionally graze it.

Seed of slender milkvetch is not commercially available. It is not used in landscaping, and it has little value for erosion control. Lakota mothers chewed the roots to improve milk flow.

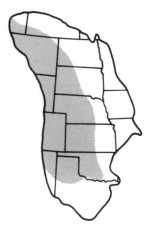

Figure 32 *Astragalus hyalinus*

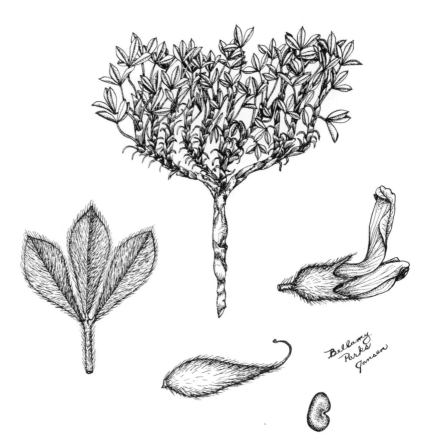

12. *Astragalus hyalinus* M. E. Jones   Transparent milkvetch (Figure 32)

[*hyalinus* (Gk.): glossy, shining, in reference to the foliage.]

Life Span: perennial. *Origin*: native. *Height*: 1–7 dm. *Stems*: very short, several to many from a caudex and woody taproot, herbaceous, low, cespitose, reduced to crowns or up to 2.5 cm long, forming mounds or cushions 1–5 dm in diameter. *Leaves*: 3-foliate, 5–45 mm long; leaflets oblanceolate or obovate, 3–15 mm long, 1–5 mm wide, apex acute or obtuse, silvery, villous-strigose, dolabriform hairs to 2.1 mm long; petioles 2–20 mm long; stipules 6–11 mm long, hyaline, connate, ciliate, transversely wrinkled. *Inflorescences*: axillary racemes, shorter than the subtending leaves, 1- to 3-flowered; peduncles 0–3.5 mm long, often hidden by stipules. *Flowers*: calyx tube cylindric, 5–7 mm long, 5-lobed; lobes 1.5–3.5 mm long, triangular to acuminate; corolla white (drying yellow), tips of wings and keel sometimes light purple; banner erect, narrow in the middle (fiddle-shaped), 1.1–1.9 cm long, claw 6–8 mm long and oblanceolate; wings 1–1.8 cm long, claws 5–8 mm long; keel 1.0–1.4 cm long, claws 6–9 mm long; all petals dorsally villous; stamens 10, diadelphous; pedicels obsolete to 0.5 mm long. *Fruit*: legumes erect, concealed by stipules, ellipsoid to ovoid, 5–7 mm long, base obtuse, with short filiform beak, densely strigulose, usually 1- or 2-seeded; seeds yellow to brown, reniform, smooth.

Synonym: *Orophaca argophylla* (Nutt.) Rydb.
Other Common Name: summer orophaca

Transparent milkvetch is rare to locally common in shallow soils of prairie hilltops and hillsides. It is unpalatable to livestock because of its low growth-form and rela-tively sharp stipules. It flowers in June and July.

Commercial seed is not available. Transparent milkvetch has little value for erosion control and has not been introduced into horticulture.

85

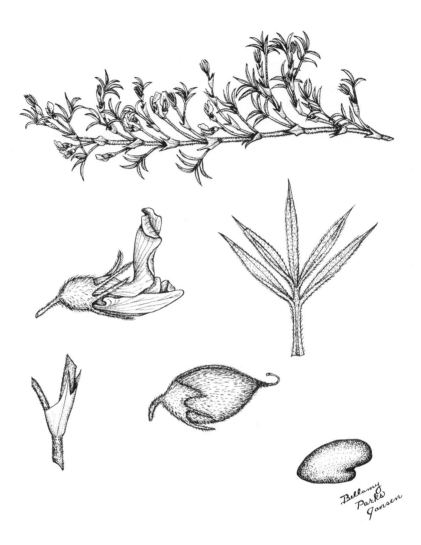

13. *Astragalus kentrophyta* A. Gray — Kentrophyta milkvetch (Figure 33)

[*kentron* (Gk.): a point or spine; + *phyton* (Gk.): plant, in reference to the spiny leaflets.]

Life Span: perennial (sometimes annual). *Origin*: native. *Height*: stems to 1.5 dm long. *Stems*: caulescent-ascending, or prostrate-pulvinate to matforming, or cespitose; mats reaching 3 dm in diameter, freely branching from a caudex and taproot. *Leaves*: pinnately to subpalmately compound, 4–25 mm long; leaflets 3–9, commonly 5, elliptic-lanceolate to subulate or oblanceolate, 3–20 mm long, 1–2 mm wide; apex with a soft to stiff spine, becoming rigid, up to 1.2 mm long; petiole 0–4 mm long; stipules connate, 1.5–6 mm long. *Inflorescences*: short axillary racemes of 1–3 flowers, these ascending, then declining; peduncles short, concealed by the stipules. *Flowers*: calyx tube campanulate to short-cylindric, 1.5–3 cm long, 5-lobed; lobes irregular, subulate, 0.5–2.5 mm long; corolla white to pinkish-purple; banner erect, 4–8 mm long; wings 4–6 mm long; keel 3–4 mm long, claws 1–2 mm long; stamens 10, diadelphous; pedicels 0.5–1.8 mm long. *Fruit*: legumes spreading or deflexed, sessile, lenticular-ovoid to ellipsoid or lanceolate-acuminate, 4–9 mm long, subterete to laterally compressed, 1.5–4 mm wide, with or without a beak, 1–3 seeded; valves thick papery, villous or strigose; seeds brown to black, smooth, often pitted, 2–3 mm long.

Synonyms: *Astragalus impensis* (Sheld.) Woot. & Standl., *A. jessiae* Peck, *A. montana* Nutt. *ex* T. & G., *A. tegetarius* S. Wats., *A. viridis* (Nutt.) Sheld., *Kentrophyta montana* Nutt., *K. viridis* Nutt.
Other Common Name: Nuttall's kentrophyta

Kentrophyta milkvetch is infrequent but widely distributed in open soil on ridges, hilltops, and bluffs. It flowers from June to August. When immature, it is fairly palatable to sheep, goats, and pronghorn. Otherwise, it is worthless.

Seed is not available from commercial sources. It has little value for landscaping or erosion control.

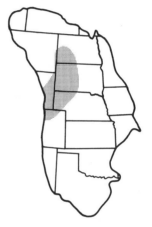

Figure 34 *Astragalus lotiflorus*

14. *Astragalus lotiflorus* Hook.

Low milkvetch (Figure 34)

[*lotus* (Lat.): name of a legume, + *floris* (Lat.): flowers, in reference to the flowers resembling those of lotus.]

Life Span: perennial (sometimes annual). **Origin**: native. **Height**:3–8(12) cm. **Stems**: herbaceous, cespitose, lower stems prostrate to ascending, upper stems erect; pubescence appressed, straight or spreading and sinuous, hairs dolabriform. **Leaves**: odd-pinnately compound, 2–14 cm long, (3)7–17 leaflets; leaflets oblong to elliptic or oblanceolate, 5–20 mm long, 1.5–7 mm wide, apex and base obtuse or acute, nearly glabrous to thinly canescent above, densely canescent beneath, hairs dolabriform; petioles 0–2 cm long; stipules somewhat membranous, lanceolate, distinct, 2–8 mm long. **Inflorescences**: terminal racemes of chasmogamous flowers, subcapitate to ovoid, 3- to 17-flowered, peduncles 3–11 cm long and seldom surpassing subtending leaves; inconspicuous cleistogamous flowers in racemes of 1–5 flowers on short peduncles near the base of the plant. **Flowers**: flowers of 2 kinds; calyx tube of chasmogamous flowers campanulate, 3–5 mm long, densely canescent, 5-lobed; lobes long-acuminate, 2–4(5) mm long; corolla greenish-white to yellowish-white, sometimes suffused with lavender or bicolored; banner strongly to moderately reflexed, 8–14 mm long; wings 7–12 mm long, claws 3–4.5 mm long; keel 6–10 mm long, claw 3.5–4.5 mm long; stamens 10, diadelphous; cleistogamous flowers 4–7 mm long. **Fruit**: legumes spreading, sessile, oblong-lanceolate to ellipsoid or ovoid-acuminate, straight or curved, 1.5–3(4) cm long, 5–8 mm in diameter, with a conspicuous beak, pubescent, unilocular, upper margin nearly straight, lower margin convex, sometimes dorsally sulcate, obcompressed or subterete, valves fleshy then coriaceous; seeds yellow to brown, sometimes with purple spots, 1.5–2.5 mm long.

Synonyms: *Batidophaca cretacea* (Buckl.) Rydb., *B. lotiflora* (Hook.) Rydb., *B. nebraskensis* (Bates) Rydb.
Other Common Name: lotus milkvetch

Low milkvetch is infrequent to common on rocky or sandy prairie hillsides and uplands, usually growing on barren or eroded sites. It flowers from March to June. It is occasionally grazed by livestock and wildlife.

Low milkvetch seed is not available from commercial sources. It has not been introduced into horticulture, and it has little value for erosion control.

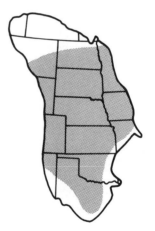

89

Figure 35 *Astragalus missouriensis*

15. *Astragalus missouriensis* Nutt.

Missouri milkvetch (Figure 35)

[*missouriensis*: of or from Missouri.]

Life Span: perennial. *Origin*: native. *Height*: 6–20 cm. *Stems*: several to many from a stout taproot, herbaceous, cespitose and spreading, lower ones prostrate or ascending, uper ones erect, densely strigose-canescent, hairs dolabriform. *Leaves*: alternate, odd-pinnately compound, 4–14 cm long; leaflets 7–19(25), oblanceolate to elliptic, 5–14(17) mm long, 2–6 mm wide, apex acute to obtuse or mucronate, densely canescent, hairs dolabriform; stipules distinct, lanceolate, 2–9 mm long. *Inflorescences*: racemes elevated above the leaves, with 3–15 flowers, scarcely elongating at maturity; peduncles 10–15 cm long, becoming prostrate in fruit. *Flowers*: calyx tube oblique, cylindric, 6–9 mm long, often dark-pigmented, strigulose, 5-lobed; lobes subulate to long acuminate, 1–4 mm long; corolla purplish-pink (rarely white or deep purple), drying blue; banner 1.5–2.5 cm long, notched, moderately to strongly reflexed; wings 1.3–2 cm long, claws 7–11 mm long, keel 1.2–1.9 cm long, claws 7–11 mm long; stamens 10, diadelphous; pedicels 2 mm long. *Fruit*: legumes spreading, oblong-cylindrical or ellipsoid, 1.5–2.5 cm long (excluding the beak), 4–10 mm wide, beak 4 mm long, glabrate to strigulose, with appressed hairs, sessile, not sulcate, nearly circular in cross-section, unilocular or subunilocular, dull-reddish, becoming black with age, becoming coriaceous; seeds brown, 2–3 mm long, surface pitted. n=11.

Synonym: *Xylophacos missouriensis* (Nutt.) Rydb.

Missouri milkvetch is common in rocky to sandy soils of prairies, dry hillsides, knolls, bluffs, washes, badlands, and stream banks. It flowers from March to July. References on forage value are conflicting. A few unsubstantiated reports indicate that it is poisonous, while others indicate high forage value.

Seeds are seldom commercially available, although it has been introduced into horticulture. Missouri milkvetch has little value for erosion control.

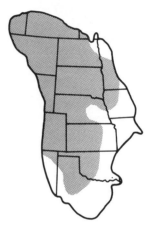

91

Figure 36   *Astragalus mollissimus*

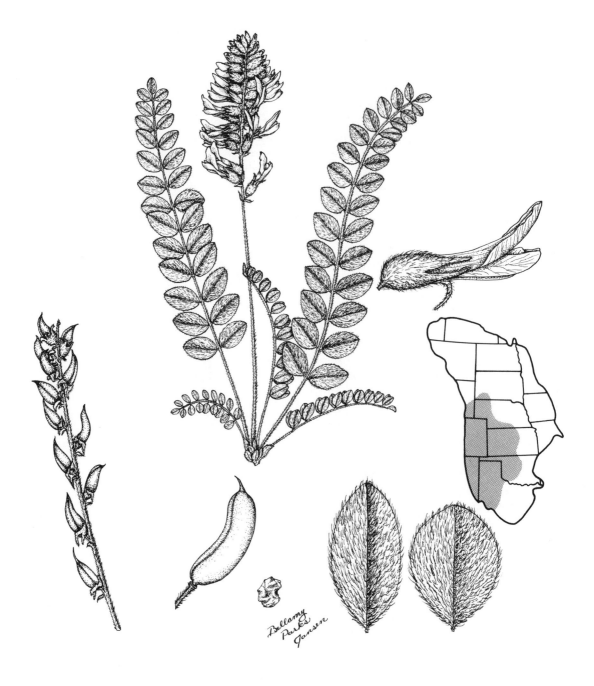

16. *Astragalus mollissimus* Torr.

Woolly locoweed (Figure 36)

[*mollissimus* (Lat.): most soft, referring to the covering of silky hairs.]

Life Span: perennial. *Origin*: native. *Height*: 2–15(25) cm. *Stems*: 1 to several from a woody taproot, herbaceous, short, stout, densely to loosely tufted, usually simple, outer ones prostrate, inner ones ascending, covered with silky-tomentose basifixed pubescence. *Leaves*: odd-pinnately compound, ascending or arching, 5–22 cm long; leaflets 15–35, obovate to oblanceolate or elliptic (infrequently lanceolate), 5–25 mm long, 2–15 mm wide, covered with long silky basifixed hairs, apex obtuse or acute, sometimes mucronate, base obtuse or acute; stipules distinct, up to 1.5 cm long, triangular, membranous, silky. *Inflorescences*: terminal racemes, oblong, 10- to 40-flowered, 4–10 cm long, longer in fruit; peduncles naked, 5–20 cm long, prostrate with maturity. *Flowers*: calyx tube cylindric, slightly oblique, 5–10 mm long, 5-lobed; lobes long-acuminate, 2–5 mm long; corolla purple to reddish-purple (rarely yellow or white), drying blue; banner 1.7–2.2 cm long, slightly to moderately reflexed; wings 1.5–2 cm long, claws 7–11 mm long; keel 1.5–1.8 cm long, claws 8–11 mm long; stamens 10, diadelphous. *Fruit*: legumes ascending or spreading, plump-ellipsoid to oblong-ellipsoid or lanceolate, 1.4–2.5 cm long including the beak, 4–9 mm wide, often broadest near the base, bilocular for most of the length, glabrous to puberulent, coriaceous to woody, many-seeded; seeds brown, 2–3 mm long, smooth to rough. 2n=22.

Synonym: *Astragalus bigelovii* A. Gray
Other Common Names: purple locoweed, stemmed loco, Texas locoweed, true loco, woolly milkvetch

Woolly locoweed is scattered to common in dry prairies, hillsides, roadsides, stream valleys, and uplands. It is most abundant in sandy or rocky soils. It flowers from April to June. It is generally unpalatable to livestock, but animals will consume the plants if other forage is not available. Some animals, especially horses, soon become addicted to the plant and refuse to eat better feed. The toxic principle is locoine, and the effect is accumulative. Symptoms of poisoning are loss of weight, depression, rough coat, and staggering gait. The optic nerve is apparently affected. Animals will leap high over small objects and shy violently from small or imaginary objects. Once animals start to walk forward, they often continue until walking into an obstruction, hence, the term "loco" which is Spanish for crazy. Both green and dried plants are poisonous.

Woolly locoweed has been introduced into horticulture, but seed is seldom commercially available. It has little value for erosion control.

Astragalus aboriginum Richards., Indian milkvetch, differs by having a bidentate apex on the wing petals, and the fruit is pendulous. It grows in the northwestern part of the Great Plains. *Astragalus shortianus* Nutt. *ex* T. & G., Waller's milkvetch, is acaulescent with strigulose fruit. It is rarely collected in the extreme western portion of the Great Plains. It is common farther west. *Astragalus purshii* Dougl. *ex* Hook., Pursh milkvetch, is also acaulescent. It has villous hairs (to 5 mm long) on the fruits and is most common in the western and northwestern Great Plains.

Figure 37   *Astragalus nuttallianus*

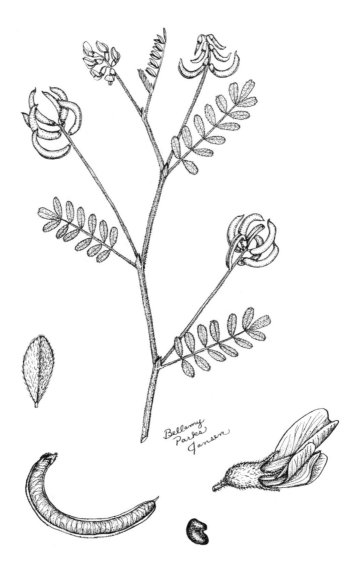

17. *Astragalus nuttallianus* DC.   Smallflowered milkvetch (Figure 37)

[*nuttallianus*: named for Thomas Nuttall (1786–1859), English-American naturalist.]

Life Span: annual or winter annual. ***Origin***: native. ***Height***: 3–40 cm. ***Stems***: herbaceous, prostrate, simple or basally branched, glabrate or silvery- strigose, from a taproot. ***Leaves***: alternate, odd-pinnately compound, 1–7 cm long, petioled or subsessile above; leaflets 7–19(25), obovate or ovate-cuneate to elliptic or oblong, 2–10(15) mm long, 1–5 mm wide, strigose, apex truncate or retuse to acute; stipules 1–6 mm, distinct. ***Inflorescences***: axillary racemes, erect or incurved-ascending, 1- to 8-flowered; peduncles 0.5–10 cm long. ***Flowers***: calyx tube 1.2–3.2 mm long, thinly or conspicuously black to white strigulose, 5-lobed; lobes 1–3 mm long; corolla white to pinkish-purple; banner 5–9 mm long, wings 5–7 mm long; keel obtuse, 4–6 mm long; stamens 10, diadelphous; pedicels commonly arched, 0.5–1 mm long in flower, 0.8–1.6 mm long in fruit. ***Fruit***: legumes spreading or declined, sessile or barely substipitate, oblong to linear-oblanceolate, incurved to nearly straight, 1–2.5 cm long, initially subtriquetrous-compressed, then subterete, 2–3.5 mm wide, glabrous or strigulose, dorsally sulcate, valves papery to thinly coriaceous, green when immature, becoming reticulate and dark brown; seeds yellow to brown, sometimes with purple spots, somewhat wrinkled, shiny.

Two varieties occur in the Great Plains. Var. *nuttallianus* has glabrous legumes, and the leaves are retuse or truncate-emarginate. It grows on mesic sites from southern Kansas to southern Texas. Var. *austrinus* (J. Small) Barneby has strigulose or glabrous legumes, and the leaves are never retuse or truncate-emarginate. It grows farther southwest, but the ranges of the two varieties overlap.

Synonym: *Astragalus austrinus* (J. Small) Schulz

Smallflowered milkvetch is scattered to common on sandy and gravelly flats, washes, and roadsides. It flowers from April to June. It has relatively low palatability. It is occasionally eaten by cattle and deer. Quail eat the seeds.

Smallflowered milkvetch has been introduced into horticulture but has little potential for landscaping. It is not valuable for erosion control.

Astragalus lindheimeri Engelm. *ex* A. Gray, Lindheimer milkvetch, is another annual or winter annual. Its banner is large, 1.2–1.9 cm long. It is found in the southern Great Plains. *Astragalus americanus* (Hook.) M. E. Jones, American milkvetch, is a perennial with distinct stipules. It is caulescent with white or ochroleucous flowers. It grows in the northwestern Great Plains.

Figure 38   *Astragalus pectinatus*

18. *Astragalus pectinatus* Dougl. *ex* G. Don   Narrowleaf poisonvetch (Figure 38)

[*pectinatus* (Lat.): comb-like, in reference to the leaves.]

Life Span: perennial. *Origin*: native. *Height*: 1–6 dm long. *Stems*: herbaceous, stout, ascending to prostrate with tips ascending, diffusely branched at or below the middle, forming mats, basally leafless, strigulose with basifixed hairs, from short rhizomes and thick taproots. *Leaves*: alternate, odd-pinnately compound, sessile, 2–10 cm long; leaflets 7–23, stiff, 1.3–7 cm long, terminal leaflet longest and continuous with the rachis, 0.5–2 mm wide, glabrate above, strigulose beneath, hairs appressed and basifixed; stipules membranous, connate, 1–10 mm long. *Inflorescences*: axillary racemes, 6- to 30-flowered, becoming loose; peduncles stout, 2–12 cm long. *Flowers*: calyx tube cylindric (6)8–10 mm long, black or white or mixed strigulose, 5-lobed; lobes triangular or short-acuminate, 1–2(3) mm long; corolla ochroleucous, drying yellow; banner strongly to moderately reflexed, 1.7–2.1 cm long, deeply notched; wings 1.7–2 cm long, claws 7–9 mm long; keel 1.5–1.7 cm long, claws 8–9 mm long; stamens 10, diadelphous; pedicels 1–4 mm long. *Fruit*: legumes eventually drooping, ovoid to ellipsoid, 1–2.5 cm long (including beak), slightly compressed, 6–8 mm wide, unilocular; valves fleshy when immature, woody when mature, glabrous, rugulose, suture prominent; seeds light or pinkish-brown, smooth or minutely pitted, 3–4 mm long.

Synonym: *Cnemidophacos pectinatus* (Hook.) Rydb.
Other Common Names: tineleaved milkvetch, tineleaved poisonvetch

Narrowleaf poisonvetch is scattered to locally abundant on dry prairie hillsides and hilltops, eroded badlands, and weedy roadsides. It is most common on clay or shale soils. It flowers in May and June. It is an indicator of selenium in the soil. It accumulates selenium and can cause poisoning in livestock. Selenium poisoning is discussed in the description of *Astragalus bisulcatus*.

Narrowleaf poisonvetch seed is not available from commercial sources. It has little potential for landscaping or erosion control.

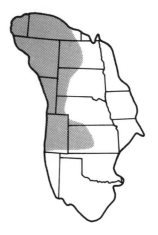

97

Figure 39   *Astragalus plattensis*

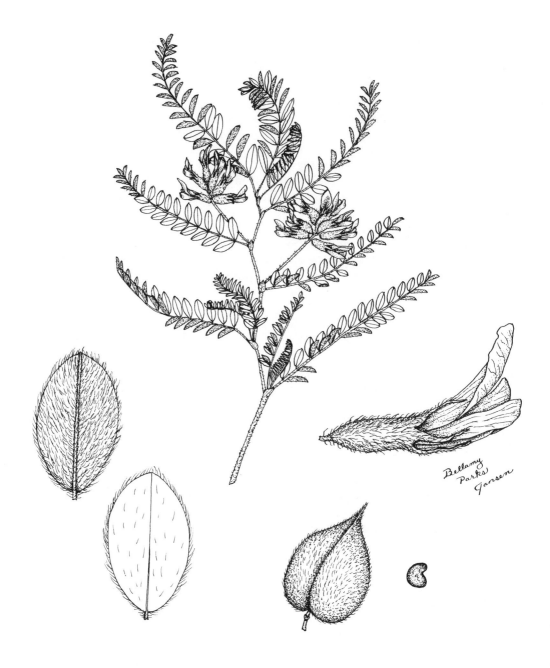

19. *Astragalus plattensis* Nutt. *ex* T. & G. Platte milkvetch (Figure 39)

[*plattensis*: of or from the Platte River region.]

Life Span: perennial. ***Origin***: native. ***Height***: 1–4 dm long. ***Stems***: usually several, often solitary, herbaceous, diffuse or spreading, decumbent, slender, villous with basifixed hairs, from elongate rhizomes and a woody taproot. ***Leaves***: odd-pinnately compound, 3–12 cm long; leaflets (9)15–35, oblong to elliptic, 8–18 mm long, 1.5–5 mm wide, apex obtuse or retuse, base obtuse to acute, glabrate above, villous with basifixed hairs beneath; short-petioled to subsessile above; stipules membranous, connate on lower nodes, distinct on upper nodes, lanceolate or ovate-acuminate. ***Inflorescences***: axillary racemes, initially subcapitate and dense with (3)6–15 flowers; peduncles stout, 4–8 cm long, usually shorter than the subtending leaves. ***Flowers***: calyx tube cylindric, slightly oblique, 6–9 mm long, white or black or mixed pubescence, 5-lobed; lobes long-acuminate, 2.5–4(6) mm long; corolla pink or pinkish-purple, drying whitish-blue; banner becoming moderately to strongly reflexed, deeply notched, 1.2–2 cm long; wings 1.2–1.8 cm long, claws 6–9 mm long; keel 1.2–1.5 cm long, claws 6–8 mm long; stamens 10, diadelphous; pedicels 2–3 mm long. ***Fruit***: legumes spreading or ascending, sessile or stipitate (stipe 0.5–1.5 mm long), ovoid or ellipsoid, pointed, 1–2 cm long (excluding the 1–3.5 mm beak), subterete or ob-compressed, 1–1.3 cm wide, fleshy, villosulous, becoming coriaceous or woody, usually bilocular; seeds pinkish-brown to black, smooth, reniform, 2.5–3.5 mm long.

Synonyms: *Astragalus pachycarpus* T. & G., *Geoprumnon plattense* (Nutt. *ex* T. & G.) Rydb.

Other Common Name: Platte River milk-vetch

Platte milkvetch is infrequent to common on rocky or sandy prairies, hillsides, woodlands, gullies, and bluffs. It flowers from March to July. It is eaten by all classes of livestock and decreases with continued heavy grazing.

Commercial seed of Platte milkvetch is generally not available. It has few applications in horticulture or for erosion control.

Astragalus neglectus (T. & G.) Sheld., Cooper milkvetch, is also caulescent, but it has erect stems and white to yellow flowers. It occurs in the northeastern Great Plains. *Astragalus praelongus* Sheld. var. *ellisiae* (Rydb.) Barneby, stinking milkvetch, is also caulescent, with erect stems, and ochroleucous flowers. These plants have strong, ill-scented odor. They grow in the extreme southwestern Great Plains.

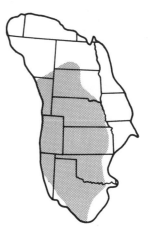

99

Figure 40 *Astragalus racemosus*

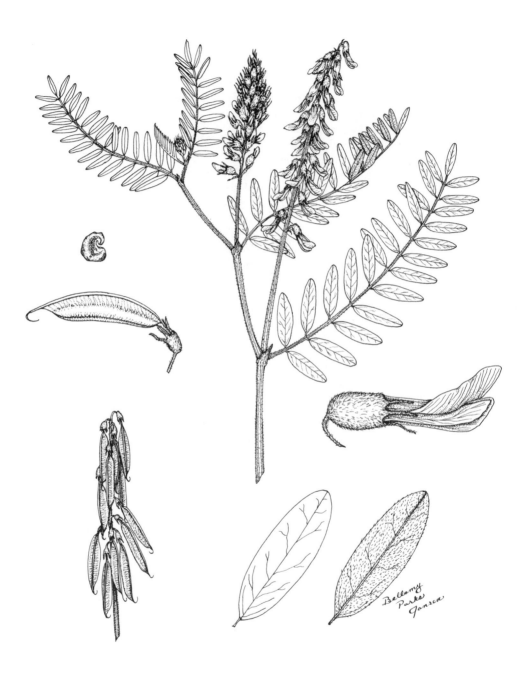

20. *Astragalus racemosus* Pursh

Racemed poisonvetch (Figure 40)

[*racemosus* (Lat.): full of clusters, in reference to the racemes.]

Life Span: perennial. *Origin*: native. *Height*: 2–7(10) dm. *Stems*: few to several from a short caudex and taproot, herbaceous, erect to ascending, usually branched above, finely strigulose with basifixed hairs. *Leaves*: alternate, odd-pinnately compound, subsessile above, short-petioled below, 4–15 cm long; leaflets 9–31, paired or irregularly arranged, linear to narrowly oblong, 1–4 cm long, 1–9 mm wide, apex obtuse to acute, sometimes mucronate, base acute, glabrous above, minutely strigose beneath, margins rolled upward when dry; stipules membranous, those of lower nodes connate, those of upper nodes distinct, 3–12 mm long. *Inflorescences*: axillary racemes, lax, 4–10 cm long, dense, 12- to 70-flowered; peduncles stout, 3–11 cm long. *Flowers*: calyx tube campanulate or cylindric, slightly oblique, 5–9 mm long, strigose, 5-lobed; lobes acuminate to subulate, 2–6 mm long; corolla white or cream, sometimes tinged with pink or purple; banner moderately to strongly reflexed, 1.3–2 cm long; wings 1.2–1.9 cm long, light purple or tipped with purple; keel 1.1–1.6 cm long, purple; stamens 10, diadelphous; pedicels in flower 2–3.5 mm long, in fruit 3–8 mm long. *Fruit*: legumes drooping, ellipsoid to linear, 1–3 cm long, triquetrous, 3–8 mm wide, acute on upper suture, widely sulcate on lower suture, stipitate (stipe 3–7 mm long), unilocular, 12- to many-seeded, valves papery, glabrous; seeds brown, often purple-spotted, somewhat shiny, 2.2–3 mm long. $n=12$.

Synonym: *Tium racemosum* (Pursh) Rydb.
Other Common Names: alkali milkvetch, creamy milkvetch, creamy poisonvetch, racemed locoweed, sunkleja hu (Lakota)

Racemed poisonvetch is infrequent to common on dry sandy or rocky prairie uplands, hillsides, eroded slopes, and disturbed roadsides. It flowers from May to July. It is an indicator of seleniferous soils and accumulates selenium. Therefore, it is a poisonous plant. Selenium poisoning is discussed under the description of *Astragalus bisulcatus*.

Seed is not commercially available. Its large flowers make it potentially useful for landscaping. Racemed poisonvetch is of little value for erosion control.

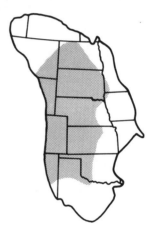

101

Figure 41   *Astragalus sericoleucus*

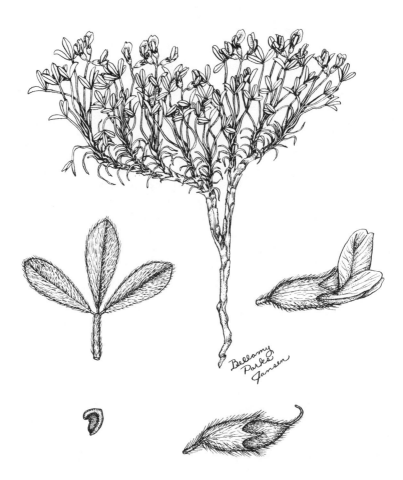

21. *Astragalus sericoleucus* A. Gray

Silky milkvetch (Figure 41)

[*serikos* (Gk.): silken; + *leukos* (Gk.): white, referring to the pubescence covering the herbage.]

Life Span: perennial. **Origin**: native. **Height**: 1–5 cm. **Stems**: herbaceous, freely branching, prostrate, forming mats, 2–15 dm in diameter; pubescence of dense, silvery, dolabriform hairs. **Leaves**: alternate, palmately 3-foliate, 1–5 cm long; leaflets broadly to narrowly oblanceolate or obovate-cuneate, 4–14 mm long, 2–7 mm wide, apex obtuse to acute; stipules connate, 2–8 mm long. **Inflorescences**: racemes (1)2- to 5-flowered, included in the stipular sheath or exserted; peduncles 5–25 mm long. **Flowers**: calyx tube campanulate, 2–3 mm long, pubescent, 5-lobed; lobes triangular to acuminate, 0.5–1.5 mm long; corolla pinkish-purple (rarely white), drying yellow; banner reflexed, 5–7 mm long, 3–4 mm wide; wings 5–6 mm long, claws 2–2.5 mm long; keel 4–4.5 mm long, claws 2–2.4 mm long; stamens 10, diadelphous; pedicels 0.5–1.5 mm long. **Fruit**: legumes ovoid-ellipsoid, tapering to a beak, 4–7 mm long, densely silky-strigose, mostly enclosed in the persistent calyx, 1- or 2-seeded; seeds greenish-brown, smooth, 1.2–1.8 mm long.

Synonyms: *Astragalus sericea* sensu aucct. non DC., *Orophaca sericea* sensu aucct. non (Nutt.) Britt.
Other Common Name: silky orophaca

Silky milkvetch is infrequent to locally common on eroded, shallow soils of hilltops and ridges. It flowers in May and June. It is unpalatable to livestock.

Seed is not available from commercial sources. It has little value for landscaping or erosion control.

Figure 42   *Astragalus spatulatus*

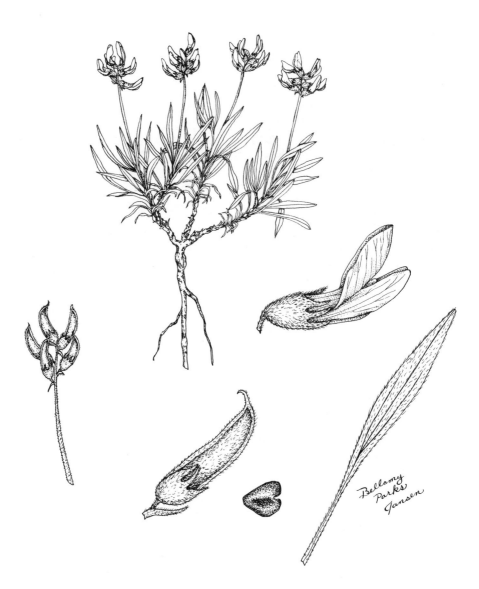

22. *Astragalus spatulatus* Sheld.

Tufted milkvetch (Figure 42)

[*spatula* (Lat.): spoon, in reference to the shape of the leaflets.]

Life Span: perennial. *Origin*: native. *Height*: 6–10 cm. *Stems*: 1 to several from a caudex and taproot, herbaceous, short, cespitose-tufted, mat-forming, covered by imbricated stipules, densely silvery-strigose, some hairs dolabriform. *Leaves*: clustered, upper leaves 1.5–6 cm long, lower leaves 0.5–1.5 cm long, mostly consisting of simple phyllodes, blades oblanceolate to linear, narrowly spatulate, acute, upper leaves sometimes with 2–4 oblong-lanceolate lateral leaflets; petioles well developed; stipules connate, membranous, 2–7 mm long. *Inflorescences*: subcapitate racemes, 2–10 cm long, 3- to 11-flowered, sometimes single-flowered; peduncles and axis elongating in fruit, 0.5–7 cm long. *Flowers*: calyx tube campanulate, with white or black pubescence, 5-lobed; lobes long-acuminate to subulate, 0.5–2.5 mm long; corolla pink to purple with white-tipped wings (rarely light yellow to purple); banner slightly to moderately reflexed, 6–10 mm long, 4.5–6.5 mm wide; wings 6–8 mm long, claws 2–3 mm long; keel 4–6 mm long, claws 2–3 mm long; stamens 10, diadelphous; pedicels 1.5 mm long in flower, 3 mm long in fruit. *Fruit*: legumes humistrate to ascending, sessile, ovoid to acuminate, slightly curved or straight, 5–16 mm long, laterally compressed, 1–4 mm wide, unilocular, valves strigose, thinly coriaceous, becoming papery; seeds purplish-brown, 2–2.5 mm long, rugulose. n=12.

Synonyms: *Astragalus caespitosus* (Nutt.) A. Gray, *Homolobus caespitosus* Nutt., *Orophaca caespitosus* (Nutt.) Britt.
Other Common Name: draba milkvetch

Tufted milkvetch is scattered to locally common in shallow or eroded soils on hilltops and bluffs. It flowers from May to June. It has low palatability to livestock. It is not important to wildlife because it is not abundant.

Tufted milkvetch has been introduced into horticulture, but seed is not commercially available. It is not valuable for erosion control.

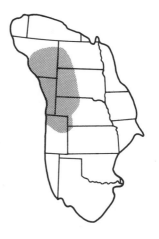

Figure 43   *Astragalus tenellus*

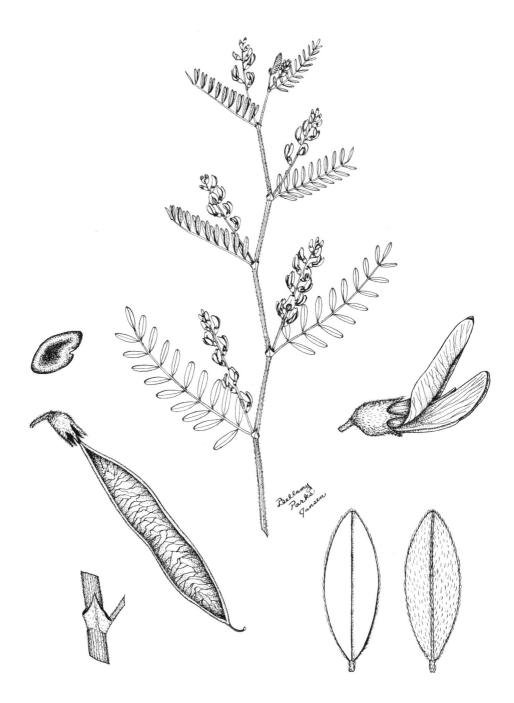

23. *Astragalus tenellus* Pursh Looseflower milkvetch (Figure 43)

[*tenellus* (Lat.): quite delicate, in reference to the leaves and flowers.]

Life Span: perennial. *Origin*: native. *Height*: 1–5 dm. *Stems*: several to many from a caudex and woody taproot, herbaceous, usually branched, thinly strigose with basifixed hairs or nearly glabrous. *Leaves*: alternate, odd-pinnately compound, 3–10 cm long, upper leaves sessile, lower leaves petiolate; leaflets (9)17–27, elliptic or linear-oblong, (3)6–20(25) mm long, apex acute or mucronate, glabrous above, thinly strigose below; stipules connate (upper stipules more distinct), deltoid, blackening on drying, 1.5–9 mm long. *Inflorescences*: axillary racemes, occasionally 2 per node, numerous, lax, 3–6 cm long, with 2–28 flowers (rarely single-flowered); peduncles 2–50 mm long. *Flowers*: calyx tube campanulate, 2–3(4) mm long, often black-hairy, 5-lobed; lobes subulate, 1–1.5(3) mm long; corolla ochroleucous to white (rarely pinkish-lavender), often with a purple-spotted keel; banner 7–10 mm long; wings 6–9 mm long, claws 2–3 mm long; keel 4–7 mm long, claws 2–3.5 mm long; stamens 10, diadelphous; pedicels 0.5–2.5 mm long in flower. *Fruit*: legumes pendulous, elliptic-oblong, 8–18 mm long, laterally compressed, 3–5 mm wide, stipitate (stipe to 7 mm long), beak 3–6 mm long, valves papery, green, mottled when immature, black at maturity, glabrate or strigose, not sulcate, unilocular; seeds brown, often-purple spotted, somewhat shiny, 2–3 mm long. n=12.

Synonyms: *Homalobus dispar* Nutt., *H. stipitatus* Rydb., *H. tenellus* (Pursh) Britt.
Other Common Names: pulse milkvetch, redstemmed milkvetch

Looseflowered milkvetch is infrequent to common in prairies, open woodlands, bluffs, gullies, badlands, and roadsides. It flowers from May to July. Palatability is variable, but it is eaten by domestic livestock and wildlife.

Commercial seed is not available. Looseflower milkvetch is not used for landscaping nor for erosion control.

Astragalus vexilliflexus Sheld., bent-flowered milkvetch, is similar, but its petals are pinkish-purple, and the legume is sessile or nearly so. It grows in the northwestern Great Plains.

107

5. BAPTISIA Vent.

[*baptisis* (Gk.): a dipping, as in dyeing, referring to some species being used as a source for a rather poor dye.]

Perennial herbs from thick rhizomes; stems herbaceous, usually much branched, the lateral branches often surpassing the central axis in length; leaves palmately 3-foliate (rarely simple), generally turning black in drying; flowers conspicuous, white, yellow, or blue to purple in color, borne in long or short racemes terminating the central axis and often also the lateral branches; calyx bilabiate, the upper lip entire, notched, or 2-lobed, the lower deeply 3-lobed; banner reniform to nearly circular, not longer than the wings, its sides usually reflexed; wings and keel nearly equal, straight, oblong, keel petals nearly separate; stamens 10, distinct; ovary stipitate; legume many-seeded, papery to woody in texture, inflated, globose to cylindric or thick-lenticular, terminating in a long or short curved beak, stalked within the persistent calyx.

About 30 species are recognized in North America. Most are located in the eastern and southern parts of the United States. Four species occur in the Great Plains.

A. Racemes very numerous, terminating the fine branches; flowers yellow; bracts small, caducous . . . . .4. *B. tinctoria*
A. Racemes 1-few, terminal, later appearing axillary by prolongation of branches; flowers white, cream-colored, or blue; bracts various, persistent or caducous
   B. Leaves, ovaries, and legumes glabrous; flowers white, blue, or tinged with purple
      C. Stipules slender, shorter than the petioles, caducous; petals white or tinged with purple; stipe 2–3 times the length of the calyx. . . . . .3. *B. lactea*
      C. Stipules broad, longer than the petioles, persistent; petals blue; stipe 1–2 times the length of the calyx. . . . . .1. *B. australis*
   B. Leaves or ovaries and legumes (at least when young) pubescent; flowers cream-colored. . . . . .2. *B. bracteata*

Figure 44 *Baptisia australis*

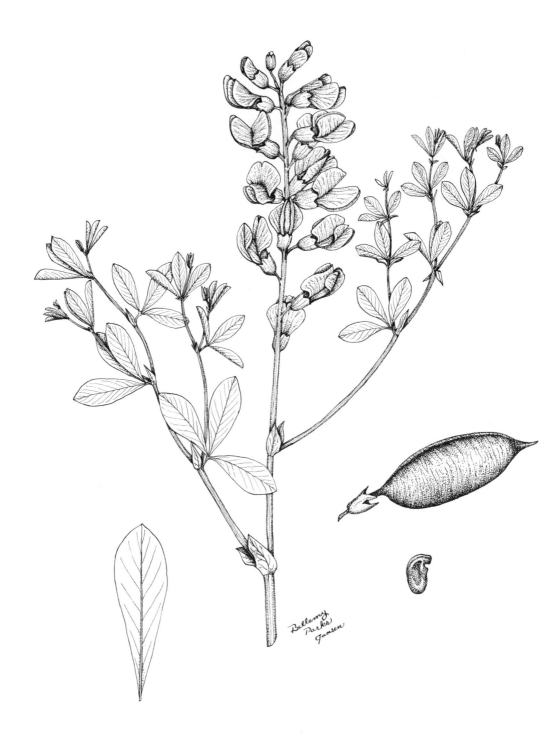

1. *Baptisia australis* (L.) R. Br.   Blue wildindigo (Figure 44)

[*australis* (Lat.): southern, in reference to its primary distribution.]

Life Span: perennial. *Origin*: native. *Height*: 4–12(15) dm. *Stems*: single primary stem with ascending or spreading branches, often from rhizomes, herbaceous, glabrous. *Leaves*: alternate, palmately 3-foliate; leaflets obovate to oblanceolate or elliptic, 1.5–3.5(8) cm long, 4–11(20) mm wide, apex rounded or acute; petioles 5–18 mm long or subsessile; stipules persistent (at least on some parts of the plants), lanceolate to ovate-lanceolate (6)8–20 mm long. *Inflorescences*: terminal raceme, erect, 1–5 dm long; bracts lanceolate, 6–15 mm long, caducous; pedicels 5–15(30) mm long. *Flowers*: calyx glabrous, 5–9(12) mm long, 4-lobed; lobes of upper lip entire or slightly notched, lobes of lower lip ovate or triangular (2–5 mm long); corolla blue to purple (rarely white), 2.5–3.5 cm long; banner 2–2.5(3) cm long, claw 2–5 mm long; wings and keel 2.5–3 cm long. *Fruit*: legume, 2–6 cm long, 1.5–2.5 cm in diameter, ellipsoid to globose, purplish-black, many-seeded, glaucous, beaked at the tip; stipe twice the length of the calyx; seeds 3.5–5 mm long, 2–2.5 mm wide, covered with resinous dots. 2n=18.

Two varieties are recognized. Var. *minor* (Lehm.) S. Wats. occurs in the Great Plains. Var. *australis* grows in the eastern and southeastern parts of the United States. Var. *minor* generally has shorter stems, more spreading branches, and smaller leaflets than var. *australis*. Also, the stipe of the fruit is about twice the length of the calyx in var. *minor* versus scarcely longer, or not at all longer, than the calyx in var. *australis*. A hybrid between *B. australis* and *B. bracteata* var. *glabrenscens* is *B.* x *bicolor* Greenm. & Larisey. It has a blue (purple) banner and yellow wings and keel. The herbage is pilosulous, and the bracts of the spike are persistent.

Synonyms: *Baptisia vespertina* J. Small, *B. minor* Lehm.
Other Common Names: blue falseindigo, rattlepod

Blue wildindigo is infrequent to common in rocky or sandy prairie, rocky open woods, limestone glades, and stream valleys. It flowers from April through June. Due to its unpleasant taste, it is seldom grazed by domestic livestock. Ingestion of relatively large quantities of the foliage may cause poisoning. Wildlife utilize some plant parts, especially the seeds.

The legumes rattle profusely when mature due to the loose seeds. The seeds are seldom viable because of insect predation. It is easy to grow from good seed and makes an attractive landscape plant.

Indians used the ripe fruits of this and other *Baptisia* as rattles. Dye was also obtained from this plant.

Figure 45   *Baptisia bracteata*

2. *Baptisia bracteata* Muhl. *ex* Ell.   Plains wildindigo (Figure 45)

[*bractea* (Lat.): covered with thin metal plates, in botany applied to the thin, scale-like leaves subtending flowers.]

Life Span: perennial. *Origin*: native. *Height*: (2)4–8 dm. *Stems*: herbaceous, coarse, ascending, with divergent branches, villous-pilosulous. *Leaves*: alternate, palmately 3-foliate; leaflets oblanceolate to elliptic, 3–8(10) cm long, 8–15 mm wide, pubescent, apex acute to obtuse; petioles 2–3(5) mm long or subsessile; stipules persistent, foliaceous, lanceolate, those of lower leaves 2–4(5) cm long, giving the appearance of 5-foliate leaves, those of upper leaves reduced. *Inflorescences*: racemes, usually solitary, sometimes declined and secund to upper side, 1–2 dm long; bracts large, foliaceous, persistent, lanceolate to oblong, 1–3 cm long, reticulately veined; pedicels 1–4 cm long. *Flowers*: calyx tube 8–11 mm, pubescent; upper lobe apically notched (nearly completely fused); 3 lower lobes deltoid, 3–4 mm long; cream-colored (white) to pale yellow, 2–3 cm long; banner 1.5–2 cm long, wing and keel 2–2.8 cm long; pistil densely hairy; ovary pubescent. *Fruit*: ovoid to ellipsoid, 2–5 cm long, black, woody, reticulately veined, pubescent, many-seeded, tapering to a long beak; stipe equaling the calyx; seeds olive to brown, 4–5 mm long, verrucose. n=9.

Two varieties are recognized. Var. *glabrescens* (Larisey) Isely is found in the Great Plains. Var. *bracteata* is found on the Coastal Plains in the southeastern United States. Unfortunately, the pubescent variety of the Great Plains takes its name from a less pubescent form and type found east of the Great Plains.

Synonym: *Baptisia leucophaea* Nutt. var. *bracteata* (Muhl. *ex* Ell.) Isley

Other Common Names: longbracted wildindigo, black rattlepod, indigo, para kari (Pawnee), tdika shande nuga (Omaha-Ponca)

Plains wildindigo is common on prairie (especially sandy and gravely soils), in dry, open woods, usually in acidic soils. It blooms from April to June. It is seldom eaten by domestic livestock because of its disagreeable taste. Consumption of large quantities may poison animals. Seeds are utilized by wildlife.

Plains wildindigo is easily grown from seed. It has escaped from cultivation in areas east of the Great Plains. It is an attractive landscape plant. Commercial seed is generally not available, and most of the wild harvested seed is damaged by insects.

Plains Indians made and consumed an infusion from the roots as a remedy for typhoid and scarlet fevers. They also consumed a decoction from leaves as a stimulant and for application to cuts and wounds. Omaha-Ponca boys used the legumes as rattles. The Pawnee pulverized the seeds, mixed the powder with bison fat, and rubbed the ointment on the abdomen for colic.

Figure 46   *Baptisia lactea*

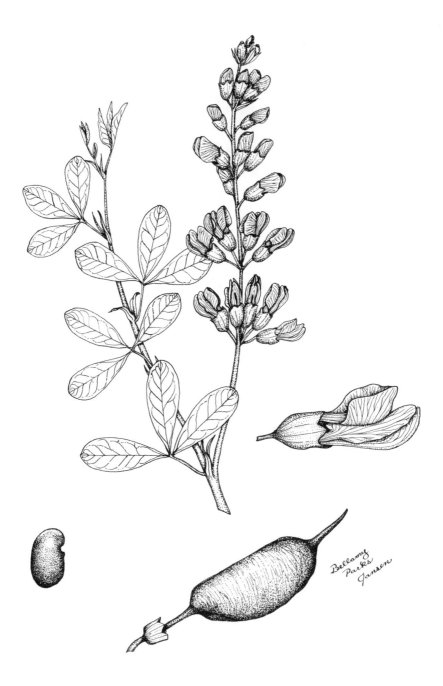

3. *Baptisia lactea* (Raf.) Thieret

White wildindigo (Figure 46)

[*lacteus* (Lat.): milky, in reference to the color of the flower of this plant.]

Life Span: perennial. *Origin*: native. *Height*: (0.5)1–1.5(2) m. *Stems*: often solitary, herbaceous, widely branched, ascending, glabrous, glaucous. *Leaves*: alternate, palmately 3-foliate, glabrous; leaflets obovate to oblanceolate, 2.5–8 cm long, 0.7–1.5(2.5) cm wide, apex obtuse to acute; petioles 6–12(25) mm long; stipules lanceolate to ovate, 5–30 mm long, caducous (some persisting until anthesis). *Inflorescences*: racemes 1 to few, erect, 2–6 dm long; bracts ovate to lanceolate, 5–14 mm long, caducous; pedicels 3–14 mm long. *Flowers*: calyx tube 8–10 mm long, densely pubescent within, persistent, 4-lobed; upper lip entire or notched; lobes of lower lip deltoid, 2–4 mm long; petals creamy-white (with a tinge of purple on the banner); banner 1–1.7 cm long; wings and keel 2–2.5 cm long. *Fruit*: legumes brownish-black, ellipsoid to oblong, glaucous, woody, 2.5–4 cm long, 8–12 mm thick, many-seeded, abruptly narrowed to a short beak; stipe 8–12 mm long, 2–3 times as long as the calyx; seeds glossy, olive to black, 3.5–5 mm long, 3 mm wide, covered with small resinous droplets. 2n=18.

Synonyms: *Baptisia leucantha* T. & G., *B. pendula* Larisey, *B. psammophila* Larisey
Other Common Names: Atlantic wildindigo, rattlepod

White wildindigo is infrequent to common in prairies, open woods, ravines, and in alluvial soils along rivers and in wet medows and valleys. It flowers from May to mid-August. Animals seldom graze white wildindigo, because it contains alkaloids that are emetic and cathartic. Young animals, particularly horses, are poisoned most frequently.

White wildindigo is easily grown from seed for landscape plantings. Commercial seed is seldom available, and insect damage to wild harvested seed is usually severe.

When plants are mowed, regrowth has more persistent stipules and larger leaves. This foliage can be confused with that of *B. australis*.

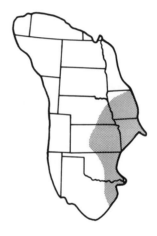

Figure 47   *Baptisia tinctoria*

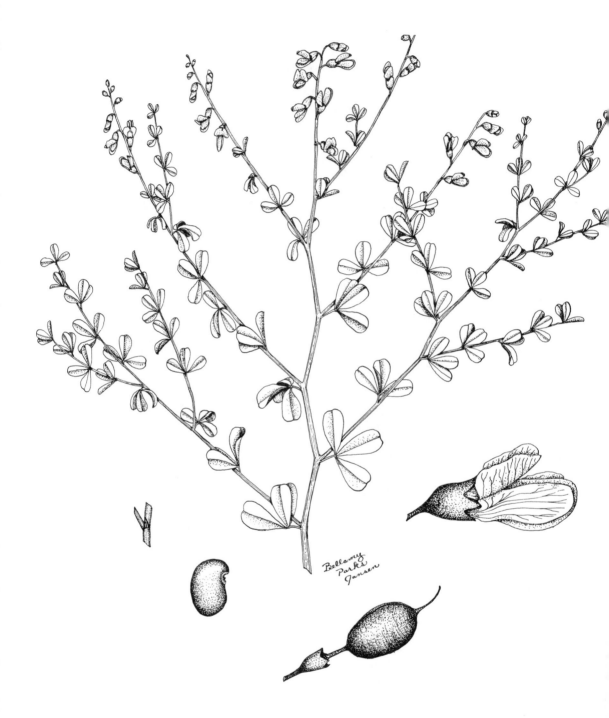

4. *Baptisia tinctoria* (L.) R. Br.　　　　　　Yellow wildindigo (Figure 47)

[*tinctorius* (Lat.): relating to colors or dyeing, in reference to dye being obtained from this plant.]

Life Span: perennial. ***Origin***: native. ***Height***: 3–10 dm. ***Stems***: herbaceous, commonly from a rhizome, with slender branches, widely spreading from the single primary stem, glabrous to slightly pubescent, glaucous at least when young, blackened in drying. ***Leaves***: alternate, palmately 3-foliate; leaflets obovate, 1–2(4) cm long, 5–12(18) mm wide, obtuse to retuse at summit, cuneate at the base, lower surface finely veined; petioles 1–5 mm long; stipules minute, setaceous, caducous. ***Inflorescences***: numerous racemes, 3–10(15) cm long, terminating most branches, loosely few-flowered; bracts small, setaceous, caducous; pedicels 3–6 mm long. ***Flowers***: calyx glabrous (margins occasionally minutely ciliate), persistent, tube 2–3 mm long, lobes on upper lip 0.5 mm long, lobes on lower lip 1–2 mm long; corolla papilionaceous, yellow, 8–13(16) mm long, wing petals 2–3 mm longer than banner; stamens 10, filaments distinct; ovary glabrous. ***Fruit***: legumes ovoid to elliptic, 8–15 mm long, 6–8 mm wide, woody, many-seeded, with a stipe 5–10 mm long; seeds yellow to light-brown, glossy. 2n=18.

Synonym: *Baptisia gibbesii* J. Small
Other Common Names: yellow falseindigo, yellow baptisia

Yellow wildindigo is scattered to common in open woods and clearings. It is occasionally found scattered in the Tallgrass Prairie. It flowers from late May through July. Yellow wildindigo is seldom eaten by domestic livestock. It apparently has a dissatisfying taste. It may be grazed by deer and other wildlife. Seeds are readily consumed by birds. It contains quinolizidine alkaloids and may be poisonous to animals.

Seed is seldom commercially available. Seeds can be harvested by hand before dehiscence. Most seed is destroyed in the legume by weevils. Germination may be improved by scarification and stratification. It is an attractive landscape plant.

Indians in the eastern Great Plains extracted dye from the foliage, although it is not the indigo dye of commerce. It appears on early lists of medicinal plants, and young shoots are sometimes eaten like asparagus. Caution must be taken, because it may be poisonous to humans.

6. CARAGANA LAM.

[*caragan* (Tr.): Mongolian name of a tree.]

Deciduous shrubs or small trees; leaves even-pinnately compound, with 2–18 small, entire leaflets; rachis often persistent and spiny; stipules small, deciduous or persistent and spiny; flowers solitary or fascicled; calyx tube campanulate or tubular, lobes nearly equal, upper 2 sometimes smaller; corolla papilionaceous, yellow (rarely white or pink); banner upright with recurved sides and long claws; wings with long claws; keel straight and obtuse; stamens 10, diadelphous; ovary sessile, rarely stipitate; legume linear, terete or inflated, usually pointed, several-seeded.

A genus of about 50 species from southern Russia to China, mostly in central Asia. Only 1 species has been introduced to the Great Plains.

Figure 48   *Caragana arborescens*

Caragana arborescens Lam.

Siberian peashrub (Figure 48)

[*arboris* (Lat.): a tree, referring to its growth form.]

Life Span: perennial. *Origin*: introduced (from Asia). *Height and Form*: shrub or small tree, 3–6 m tall. *Twigs*: several main branches, twigs remaining green for several years, young branches pubescent. *Leaves*: alternate or often fascicled, petiolate, even-pinnately compound with 8–14 leaflets; leaflets 1–2.5(3) cm long, oblong-elliptic to ovate, apex obtuse to truncate, mucronate, base obtuse, margins entire, pubescent then becoming glabrescent; rachis ending in a short spine; stipules 5–9 mm long, linear, tending toward spininess. *Inflorescences*: fascicles of (1)2–4 flowers. *Flowers*: calyx tube 5–6 mm long, minutely pubescent, 5-lobed; lobes 1 mm long; corolla bright yellow; banner 1.8–2.1 cm long, 1.7–1.9 cm wide, notched apically, narrowed abruptly to a claw 4–6 mm long; wings 1.8–2.1 cm long; keel 1.5–1.8 cm long; stamens 10, diadelphous; pedicels 1–4.5 cm long. *Fruit*: legumes linear, (2)3–5 cm long, 5 mm wide, compressed, sessile, narrowed to a slender beak, several-seeded; seeds dark brown, smooth, lustrous, 4–5 mm long, 2.5–3 mm wide.

Other Common Names: common caragana, Siberian peatree

Siberian peashrub has escaped from shelterbelts to pastures, abandoned fields, roadsides, and waste places. It grows best on sandy and slightly alkaline soils. It flowers in May and June. Palatability to domestic livestock is relatively low. It provides woody cover for wildlife.

Plants of several cultivars are available from nurseries. It is easily propagated and tolerant of extremely cold temperatures. It is most commonly planted as a windbreak border. It is also placed in buffer strips and in hedgerows. Siberian peashrub is occasionally used as an ornamental.

7. CORONILLA L.

[*coronula* (Lat.): diminutive of crown, in reference to the inflorescences.]

Herbs or shrubs with odd-pinnately compound leaves, glabrous throughout; flowers pink, white, or yellow, in long-peduncled axillary umbels; calyx tube bilabiate, broader lower lip with 3 short triangular teeth, upper lip narrow, shallowly cleft; petals about equal in length, clawed; keel petals curved upward; stamens 10, diadelphous; legumes terete or 4-angled, straight or curved, transversely jointed, few- to several-seeded.

A genus of about 25 species. All are native to Eurasia and northern Africa. Only 1 is found in the Great Plains.

Figure 49   *Coronilla varia*

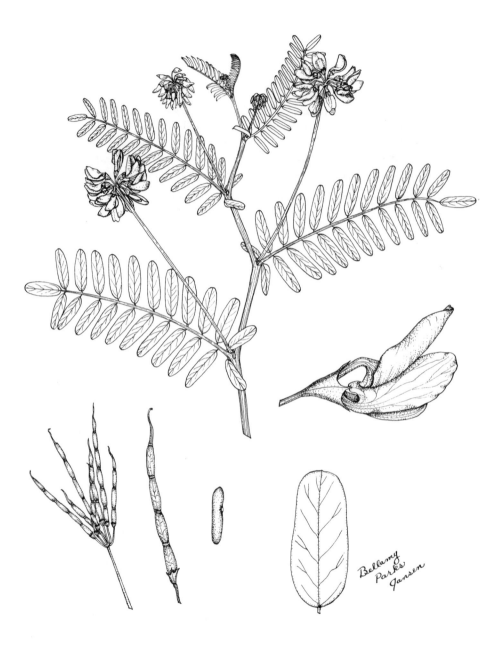

Coronilla varia L.   Crownvetch (Figure 49)

[*varius* (Lat.): different, changing, variable.]

Life Span: perennial. *Origin*: introduced (from Europe). *Height*: 3–5 dm. *Stems*: slender, ascending, glabrous, from taproots and rhizomes. *Leaves*: odd-pinnately compound, glabrous, short-petiolate below, nearly sessile above; leaflets 9–25, oblong to obovate, 1–3 cm long, acute to rounded or retuse at the summit, broadly obtuse to cuneate at the base; stipules persistent, 2–3 mm long, narrowly lanceolate. *Inflorescences*: umbels, globose, containing 8–15(20) flowers; peduncles stout, equaling or surpassing the subtending leaves, 4–12 cm long; pedicels 3–7 mm long. *Flowers*: calyx tube 1–2.5 mm long, campanulate to hemispheric, 5-lobed; upper 2 lobes nearly united, lower 3 distinct; corolla pink (white to purple); banner 10–13 mm long, claw 1–2 mm long; wings 11–15 mm long, claw 2–4 mm long; keel curved, tipped with purple, 10–13 mm long, claw 3 mm long. *Fruit*: loments, coriaceous, linear, 2–4 cm long, 3–7(12) joints, indehiscent; seeds cylindrical, slightly flattened, 3–4 mm long, 1 mm wide, dark brown, smooth. n=12.

Crownvetch is found on stream banks, roadsides, and embankments. It also grows in fields and gardens. It is usually found close to where it was originally planted, as it does not readily migrate. It flowers from May to August. It produces satisfactory forage for livestock. Grazing reduces its persistence more than does frequent clipping.

Crownvetch does not cause bloat, but it contains a poisonous nitroglycoside called coronarian. It is not toxic to ruminants because coronarian is detoxified in the rumen. Poisoned nonruminants exhibit growth depression, ataxia, and posterior paralysis. They may die. It has been known for over 150 years to be poisonous.

One of its most valuable uses is as a bank stabilizer. Crownvetch grows well in poor soils. It has been used as an understory on Christmas tree farms. It does not seed well. Seeds must be scarified and inoculated with rhizobial bacteria. It is easily transplanted, and crowns are readily available commercially. Crownvetch makes an excellent groundcover for ornamental purposes. It has been used as a landscape plant in Europe for over a century.

8. CROTALARIA L.

[*krotalon* (Gk.): a rattle, in reference to inflated pods with loose seeds after ripening.]

Annual or perennial herbs, shrubs in the tropics; leaves simple or 3-foliate; flowers in racemes, yellow; corolla not exceeding the calyx lobes; banner suborbicular, short clawed; wings not articulate at the base; keel connivent on both margins; stamens 10, monadelphous below the middle; ovary typically sessile; calyx obscurely bilabiate, upper lip less deeply cleft; legumes subglobose to cylindric or ellipsoid, greatly inflated, several- to many-seeded.

A genus of over 250 species. Most are tropical in both hemispheres. One species is reported in the Great Plains. It may have been brought into the Great Plains in hay from the southeast.

Figure 50   *Crotalaria sagittalis*

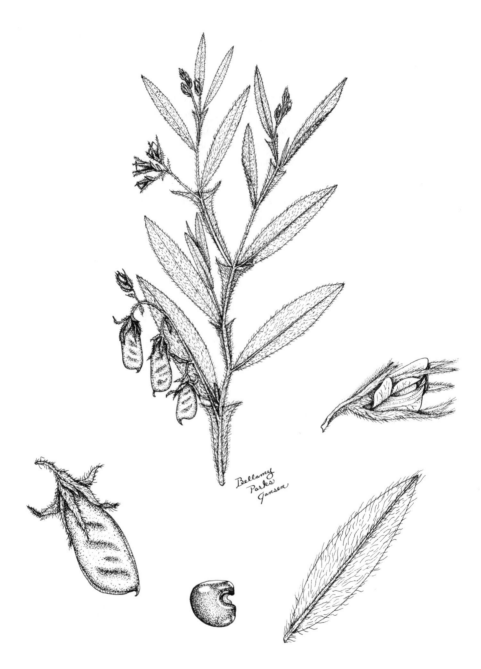

Crotalaria sagittalis L.

Rattlebox (Figure 50)

[*sagitta* (Lat.): an arrow, in reference to the shape of the stipules on the upper leaves.]

Life Span: annual in the Great Plains, but sometimes a short-lived perennial in southern states. *Origin*: native. *Height*: 1–4 dm. *Stems*: herbaceous, erect or ascending, simple or branched above, villous. *Leaves*: simple, sessile or nearly so, villous, upper leaves lanceolate to linear, (1.5)3–8 cm long, 6–15 mm wide; basal leaves smaller, elliptic to oval; stipules persistent (at least on upper leaves), inversely sagittate, decurrent on the stem for 1/2 or more of the internode, 2–3 mm wide. *Inflorescences*: 2- to 4-flowered racemes, terminating stems and branches; peduncle 1–5 cm long, pedicels (1)4–10 mm long. *Flowers*: calyx tube villous, 7–12(20) mm long; lobes 5, linear to lanceolate, unequal, upper lobe the largest, 8–10 mm long; corolla papilionaceous, yellow, petals shorter than the calyx, banner 8 mm long; stamens 10, 5 shorter than the others; ovary glabrous. *Fruit*: legume, oblong, 2–3(4) cm long, 1 cm thick, dull-brown, nearly black when ripe, greatly inflated, glabrous, sessile in calyx, many-seeded; seeds obliquely reniform, gray-brown, glossy to lustrous, 2.5–3 mm broad. n=16.

Other Common Names: arrow rattlebox, arrow crotalaria

Rattlebox is infrequent to locally common on dry, sandy, or gravelly soils on disturbed sites, wasteland, prairie glades, prairies, and open wooded slopes. It flowers from May to September. Although rattlebox is native to North America, it is not native to the Great Plains. It apparently was introduced into the Great Plains in hay shipped from the southeast.

Generally livestock avoid rattlebox, but they may develop a taste for it. It has been known for more than a century to be a poisonous plant. The poisoning was called "Missouri Bottom Disease" because horses grazing the Missouri River bottoms were affected while those grazing the adjacent uplands were not poisoned. The poisonous principle is unknown, although it is thought to be an alkaloid. Seeds are the most poisonous plant part, but the poison is present in fresh or dried herbage. Horses are more readily poisoned than cattle. Symptoms of poisoning include slow emaciation, weakness, and stupor. It may take several weeks or even a few months before the animal dies from degeneration of the liver and spleen. Birds, especially quail, consume the seed with no ill effect.

Rattlebox has no landscaping value. Seed is not commercially available. Seed requires scarification before water can be imbibed for germination.

129

9. DALEA Lucanus

[*Dalea*: named after Samuel Dale (1659–1739), English botanist.]

Annual or perennial herbs or shrubs, stems leafy; leaves alternate, small, odd-pinnately compound or 3-foliate, gland-dotted; leaflets entire; stipules subglandular or herbaceous; inflorescences of terminal spikes; flowers of various colors, small, subtended by a deciduous or persistent bract, usually gland-dotted; calyx tube campanulate or tubular, 10-ribbed, 5-lobed; lobes about equal; corolla papilionaceous or obscurely so; stamens 5–10, monadelphous; legumes indehiscent, included in persistent calyx or partially extended, thin-walled, usually somewhat compressed, 1-seeded.

About 250 species of *Dalea* have been described. They are most abundant in Mexico and the southern and southwestern United States. Fourteen occur in the Great Plains, 10 of which may be considered common or abundant.

A. Plants shrubs. . . . . .5. *D. formosa*
A. Plants annuals or herbaceous perennials
    B. Plants annual. . . . . .7. *D. leporina*
    B. Plants perennial
        C. Stems prostrate . . . . .6. *D. lanata*
        C. Stems erect or ascending
            D. Stems and leaves mainly glabrous
                E. Spike sparingly flowered . . . . .4. *D. enneandra*
                E. Spike densely flowered
                    F. Petals purple or rose. . . . . .9. *D. purpurea*
                    F. Petals yellowish-white or white
                        G. Spikes subglobose, 1.5 cm long or less. . . . . .8. *D. multiflora*
                        G. Spikes cylindric, over 1.5 cm long
                            H. Spikes mostly under 6 cm long; calyx tube glabrous or with hairs under 0.5 mm long . . . . .2. *D. candida*
                            H. Spikes mostly over 6 cm long; calyx tube densely pilose with hairs over 1.5 mm long . . . . .3. *D. cylindriceps*
            D. Stems and leaves softly villous-tomentose to silky pilose
                I. Calyx lobes greater than 2.2 mm long. .1. *D. aurea*
                I. Calyx lobes less than 2.2 mm long. .10. *D. villosa*

Figure 51   *Dalea aurea*

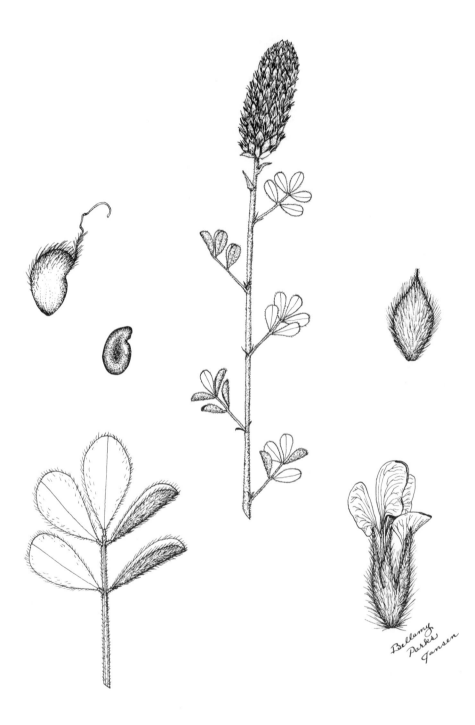

1. *Dalea aurea* Nutt. *ex* Pursh

Silktop dalea (Figure 51)

[*aureus* (Lat.): golden, in reference to the flower color.]

Life Span: perennial. *Origin*: native. *Height*: 2–7 dm. *Stems*: herbaceous, erect or ascending from a short caudex and deep taproot; stout, simple or branched above, with silky-appressed pubescence. *Leaves*: alternate, odd-pinnately compound, 1.5–4 cm long, with 3 to 9 leaflets, commonly with 5 leaflets; leaflets elliptic to oblanceolate or obovate, 5–20 mm long, 2–8 mm wide, apex obtuse or rarely acute, base acute, silky-appressed pubescence mainly beneath; petiole 3–15 mm long; stipules subulate. *Inflorescences*: terminal spikes, densely flowered, cylindric, the axis pilosulous. *Flowers*: yellow, remaining yellow on drying; calyx tube turbinate, persistent, 2.2–3 mm long, silky; lobes 5, long acuminate, 3.5–5 mm long; banner sagittate, 6–9 mm long, short-clawed; wings oblong, 5–6 mm long; keel oblong, 5–9 mm long, stamens 10 (sometimes 9). *Fruit*: legume, upper portion silky-villous, glabrous (or nearly so) below, 1-seeded; seeds broadly ellipsoid to reniform, yellow to dark brown, smooth. 2n=14.

Synonym: *Parosela aurea* (Nutt.) Britt.
Other Common Names: golden dalea, golden prairieclover, pezhuta pa (Lakota)

Silktop dalea is infrequent to locally common on prairies, open woods, brushy hillsides, and ravines. It flowers from June to September. It is readily eaten by domestic livestock and wildlife. Seeds are utilized by birds and rodents.

Seeds are infrequently available from commercial sources. Scarification improves germination. Silktop dalea has a potential for use in landscapes.

Lakota Indians made and drank a de-

octon of leaves for colic and dysentery.

Dalea nana Torr. *ex* A. Gray, dwarf dalea, is similar in appearance, except that *D. nana* is seldom taller than 3.5 dm. Its stems are diffusely branched, and its yellow petals fade to pink or brown on drying. Dwarf dalea grows in the southwestern Great Plains.

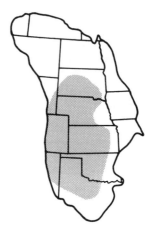

Figure 52   *Dalea candida*

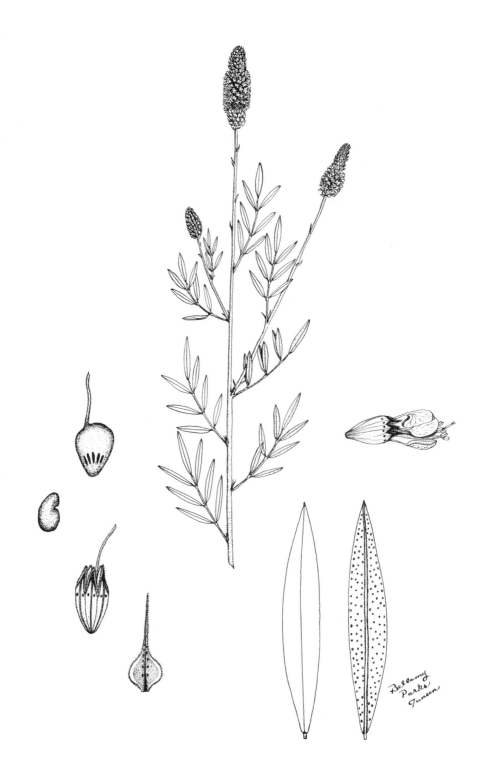

2. *Dalea candida* Michx. *ex* Willd.

White prairieclover (Figure 52)

[*candidus* (Lat.): dazzling white, in reference to the flower color.]

Life Span: perennial. *Origin*: native. *Height*: 3–7(10) dm. *Stems*: herbaceous, simple or branched above, glabrous, erect, 1 to several from a woody caudex and taproot. *Leaves*: alternate, odd-pinnately compound, 2–6 cm long; leaflets 5–9(13) (commonly 7), 1–3 cm long, 2–6 mm wide, narrowly oblanceolate to narrowly elliptic or linear, apex acute or obtuse, often mucronate, glandular-punctate beneath, often folded along the midrib; stipules glabrous, subulate. *Inflorescences*: 1 to few, spikes densely flowered, terminal, cylindric to ovoid, 2–5(8) cm long, 7–9 mm wide; bracts oblanceolate, tapering to acuminate, glabrous, gland-dotted, deciduous, ciliate, the slender tip surpassing the calyx lobes. *Flowers*: white; calyx tube 2–3 mm long, 10-ribbed, ring of glands near the top, glabrous to short-villous; lobes 5, 0.5–1.8 mm long, finely ciliate; banner 4–6 mm long; wings and keel 3–5.5 mm long; stamens 5. *Fruit*: legume, thin-walled, 2.5–4.5 mm long, usually exserted, glandular, 1-seeded; seeds brown, smooth, 1.5–2 mm long, asymmetrically reniform. 2n=14.

Two varieties are recognized. Var. *candida* has dense and conelike spikes, and its calyx tubes are glabrous. It occurs most frequently in the eastern Great Plains. Var. *oligophylla* (Torr.) Shinners has lax spikes, the axis is partially visible, and the calyx tube is pubescent. It is found in the western Great Plains.

Synonyms: *Petalostemon candidus* (Willd.) Michx., *P. occidentale* (Heller) Fern., *P. oligophyllus* Torr.
Other Common Name: bloka (Lakota)

White prairieclover is common in dry prairies and rocky upland woods. It flowers from June through July. It is palatable to all classes of livestock and adds quality to prairie hay. It decreases with continued heavy grazing. Deer, elk, pronghorn, and eastern turkey eat the foliage. Plains pocket gophers utilize the taproots. Numerous birds and rodents eat the seed.

White prairieclover usually has an excellent seed set. Germination can be improved with scarification. Branch tip cuttings readily root in a mist bench. Seed is readily available commercially. It has the potential for soil stabilization and is an attractive landscape plant.

Lakota chewed the roots for their pleasant taste, and they made tea from the leaves. Other Great Plains Indians bruised the leaves and steeped them in water for application to fresh wounds.

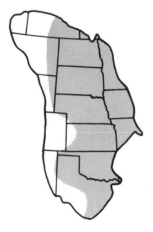

Figure 53 *Dalea cylindriceps*

Bellamy
Parks
Jansen

3. *Dalea cylindriceps* Barneby

Largespike prairieclover (Figure 53)

[*kylindros* (Gk.): a cylinder, in reference to the shape of the inflorescence.]

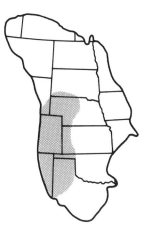

Life Span: perennial. **Origin**: native. **Height**: 2–6(8) dm. **Stems**: 1 to several from a woody caudex and taproot, herbaceous, striated, glabrous to the inflorescence, simple or few-branched. **Leaves**: alternate, odd-pinnately compound, 3–8 cm long; leaflets 5–11, 1–2.5 cm long, elliptic-oblong, glabrous, glandular below; stipules lanceolate to subulate. **Inflorescences**: terminal spikes, dense, (4)7–15(24) cm long, 8–12 mm wide, cylindric, frequently recurved at the tip, long peduncled; bracts long-attenuate, longer than the calyx, silky-villous, apex dark. **Flowers**: white to light yellow (rarely light pink), fading to light yellow; calyx tube campanulate, densely pilose, 10-ribbed; lobes 5, 1.5–2.5 mm long, deltoid-acuminate; banner 5–6 mm long, claw 3–4 mm long; wings and keel 2–5 mm long; stamens 5. **Fruit**: legume, 2.5–3 mm long, not exserted, valves pilose, 1-seeded; seed olive to yellowish-brown, smooth, beaked, 1.5–2.5 mm long. 2n=24.

Synonym: *Petalostemon compactus* (Spreng.) Swezey
Other Common Names: massivespike prairieclover, denseflowered prairieclover

Largespike prairieclover is rare to scattered in sandy and gravelly well-drained soils of prairies and stream valleys. It flowers from May to September. It is grazed by cattle and horses, as well as by wildlife. It decreases with continued heavy grazing.

Largespike prairieclover is seldom collected, and little is known about its life cycle or ecology. It is attractive and, therefore, has the potential for use in landscapes. Seed is not commercially available.

Figure 54   *Dalea enneandra*

4. *Dalea enneandra* Nutt.

Slender dalea (Figure 54)

[*ennea* (Gk.): nine; + *andros* (Gk.): man, referring to 9 stamens.]

Life Span: perennial. *Origin*: native. *Height*: 4–10 dm. *Stems*: 1–3 from a woody caudex and taproot, herbaceous, unbranched below, with many spreading branches above, glabrous, gland-dotted. *Leaves*: alternate, odd-pinnately compound, subsessile, 1–2.5 cm long; leaflets 5–13, commonly 9, linear to narrowly oblong, 3–12 mm long, 0.5–1.5 mm wide, glandular-punctate beneath, sometimes involute. *Inflorescences*: terminal spikes, loose and open, 3–10 cm long, 5- to 25-flowered, appearing 2-ranked; bracts subrotund, cuspidate, 3–4 mm long, persistent, gland-dotted, with a prominent white membranous margin. *Flowers*: white; calyx tube 3–4 mm long, silky-pilosulous; lobes 5, setaceous above, triangular below, 3.5–4.5 mm long, longer than the tube, plumose, persistent; banner 6–8 mm long, claw 2.5–3.5 mm long; wings 3–5 mm long; keel 9–12 mm long; stamens 9. *Fruit*: legume about equal to the calyx tube, 3–4 mm long, upper portion pubescent, glabrous below, 1-seeded; seeds 2.5 mm long, narrowly ovoid to reniform, beaked, yellow to brown, smooth. 2n=24.

Synonyms: *Parosela enneandra* (Nutt.) Britt., *Dalea laxiflora* Pursh
Other Common Names: nineanther dalea, plume dalea

Slender dalea is infrequent to locally abundant in dry prairies, stream valleys, and roadsides. It is most common in calcareous, rocky, or sandy soils. It flowers from June to September. It is palatable to livestock and wildlife. Seeds are eaten by birds and rodents.

Slender dalea seed is not commercially available. It has only limited potential for soil stabilization and landscaping.

Lakota Indians reported a narcotic or poisonous effect from the root. This report is unsubstantiated.

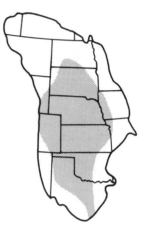

Figure 55   *Dalea formosa*

5. *Dalea formosa* Torr.   Feather plume (Figure 55)

[*formosus* (Lat.): beautiful, in reference to the flower.]

Life Span: perennial. **Origin**: native. **Height and Form**: low shrub, 1.5–10 dm tall. **Stems**: woody branches glabrous, young twigs gland-dotted. **Leaves**: alternate, odd-pinnately compound, 3–12(15) mm long, glabrous; leaflets 5–15, oblanceolate to narrowly obovate or obovate-cuneate, 1–7 mm long, often folded, gland-dotted beneath. **Inflorescences**: terminal spikes on branches and spurs, sub-capitate, 2- to 9-flowered. **Flowers**: calyx tube 3–5 mm long, pilose, with hairs 1–3 mm long; lobes 5, 4–10 mm long, filiform, plumose; banner yellow changing to rose-purple, 7–9 mm long, claw 4–5 mm long; wings 8–10 mm long, rose-purple; keel 8–12 mm long, rose-purple; stamens 10. **Fruit**: legume, glabrous to pilose above, margin pubescent, glandular, 3–3.5 mm long, 1-seeded; seeds yellow to brown, smooth, 3 mm long. $2n=14$.

Feather plume is infrequent to common on prairie hills, shrubby barren areas, and rocky hillsides. It flowers from June to September. It is valuable browse for livestock and deer.

Feather plume seed is not commercially available. It has limited potential for erosion control and landscaping.

Dalea frutescens A. Gray, black dalea, is another shrub. Its calyx lobes are much shorter than the tube, flowers are purple, and the leaves are 1–2 cm long. The ranges of these two shrubs overlap in the southern and southwestern Great Plains.

Figure 56   *Dalea lanata*

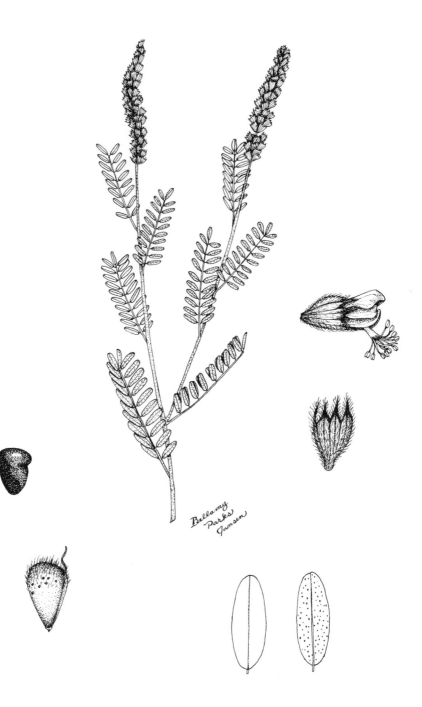

6. *Dalea lanata* Spreng.

Woolly dalea (Figure 56)

[*lanatus* (Lat.): woolly, describing the pubescence on the leaves and fruit.]

Life Span: perennial. *Origin*: native. *Height*: prostrate, 2–5(10) dm long. *Stems*: 1 to several from a caudex and taproot, herbaceous, prostrate, divaricately branching from the base, occasionally forming mats, short villous-tomentose. *Leaves*: alternate, odd-pinnately compound, 1.5–3 cm long; leaflets 15–31, obovate to oblanceolate, (3)5–10 mm long, 1.5–6 mm wide, apex obtuse, densely short-villous, gland-dotted below, short-petioled or subsessile, subtended by a gland on the rachis; stipules 1–2.5 mm long. *Inflorescences*: terminal spikes, 2–9 cm long, narrow; bracts subtending flowers persistent, gland-dotted, obovate; peduncles 4–40 mm long. *Flowers*: red to purple; calyx tube 2–2.5 mm long, gland-dotted, villosulous; lobes 5, lanceolate to triangular, 1.5–2.5 mm long; petals with a few scattered glandular dots; banner cordate and somewhat head-like, 3–4 mm long; wings 2–4 mm long, short-clawed; keel 3.5 mm long; stamens 8–10. *Fruit*: legume, gland-dotted, 2.5–3 mm long, pilosulous, 1-seeded; seeds orangish-brown to gray, beaked, shiny, slightly flattened, 2 mm long. 2n=14.

Synonyms: *Parosela lanata* (Spreng.) Britt., *Dalea glaberrima* S. Wats.

Woolly dalea is infrequent to common in sandy soils of flood plains, prairie, and roadsides. It flowers from July to October. It is eaten by domestic livestock and wildlife. Birds and rodents eat the seeds.

Commercial seed of woolly dalea is not available. Seeds are commonly eaten by insects before ripening. It has little value for landscaping, but it is a potential soil stabilizer.

Dalea jamesii (Torr.) T. & G., James' dalea, is a similar species. Its stems are erect (1–12 cm tall), leaves are mainly 3-foliate, and it has yellow flowers. James' dalea is found in the southwestern Great Plains.

Figure 57 *Dalea leporina*

7. *Dalea leporina* (Ait.) Bullock

Foxtail dalea (Figure 57)

[*leporinus* (Lat.): of hares, in reference to the silky inflorescence which resembles a rabbit's foot.]

Life Span: annual. *Origin*: native. *Height*: 2–10 dm. *Stems*: herbaceous, erect, glabrous, gland-dotted, branching from midstem or near the base. *Leaves*: alternate, odd-pinnately compound, 3–6(10) cm long; leaflets 13–49, elliptic to lanceolate, 3–12 mm long, 1–3 mm wide, gland-dotted, apex obtuse, base acute, each subtended by a gland; petioles 0–8 mm long; stipules 1–3 mm long, lanceolate to subulate. *Inflorescences*: erect spikes, terminal, dense, 1.5–8 cm long, 8–10 mm wide, cylindric; bracts subtending the flowers obovate, tip long-acuminate; peduncles 3–9 cm long. *Flowers*: pink to white, tinged with blue; calyx tube 2–2.5 mm long, gland-dotted, villous; lobes 5, 1–2 mm long, shorter than the tube, lanceolate-acuminate; banner 4–5 mm long; wings 2–3 mm long, keel 1.5–3 mm long; stamens 9 or 10. *Fruit*: legume, 2.5–3 mm long, gland-dotted, obovoid, papery, pubescent above, 1-seeded, style beak positioned at one side; seeds gray to brown, 2–2.5 mm long, smooth. 2n=14.

Synonyms: *Dalea alopecurioides* Willd., *D. lagopus* (Cav.) Willd., *Parosela alopecurioides* (Willd.) Rydb.
Other Common Name: hare's-foot dalea

Foxtail dalea is scattered to common in moist, alluvial, sandy soil of disturbed sites, roadsides, edges of fields, wooded areas, and stream banks. It flowers from July to September. It is grazed by livestock, but there is no record of use by wildlife.

Seed of foxtail dalea is occasionally commercially available. It has been seeded and plowed under as a green manure crop. It has a limited value for erosion control and little landscaping potential.

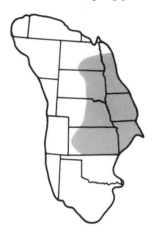

145

Figure 58   *Dalea multiflora*

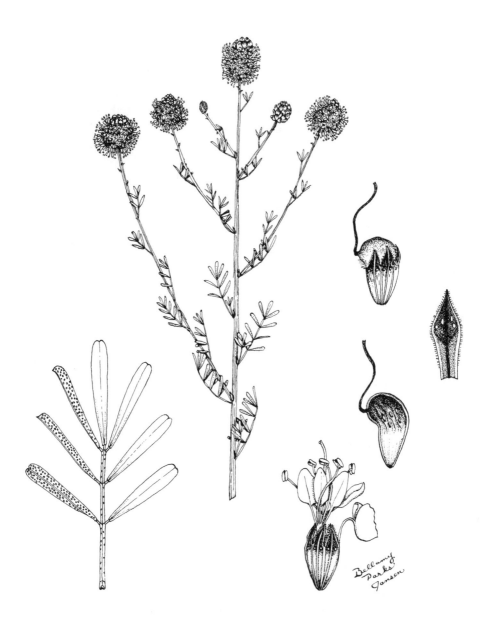

8. *Dalea multiflora* (Nutt.) Shinners　　　Roundhead prairieclover (Figure 58)

[*multus* (Lat.): many; + *floris* (Lat.): flower, in reference to the numerous spikes of flowers.]

Life Span: perennial. *Origin*: native. *Height*: 2–8 dm. *Stems*: herbaceous, much branched, usually naked below, glabrous, erect, several from a woody caudex and taproot. *Leaves*: alternate, odd-pinnately compound, 1.5–4 cm long; leaflets 3–13, commonly 7 or 9, linear to narrowly oblong, 5–15 mm long, 1–2 mm wide, obtuse to acute, sometimes mucronate, glandular-punctate beneath, often folded at the midrib, involute when dry; stipules subulate to lanceolate, 1–2 mm long. *Inflorescences*: numerous subglobose to short-ovoid spikes, 1–1.5 cm long; bracts glabrous, broad, tapering to a short acuminate tip, shorter than the calyx, caducous. *Flowers*: dense and overlapping, white; calyx tube 2.5 mm long, 10-ribbed, glabrous to finely pubescent; lobes 5, deltoid, 1–1.5 mm long, dark, ciliate; banner 5–6 mm long, claw 2–3 mm long; wings and keel 3.5–5 mm long; stamens 5. *Fruit*: legume, obliquely obovoid, 3–5 mm long, partially exserted, 1-seeded; seeds brown or tan, smooth, 2–2.5 mm long, asymmetrically reniform. 2n=14.

Synonym: *Petalostemon multiflorus* Nutt.

Roundhead prairieclover is infrequent to common on dry hills, prairies, brushy hillsides, and roadsides. It is most abundant on rocky, limestone soils. It flowers from May to August. It is grazed by all classes of domestic livestock and by deer.

Small quantities of seed are sometimes available commercially. Scarification improves germination. It has limited potential as a soil stabilizer. Its numerous spikes make it an attractive landscape plant.

Great Plains Indians made tea from leaves to be used as a preventive medicine. They chewed the roots for their pleasant taste, and stems were tied together for brooms.

In comparison to white prairieclover (*Dalea candida*), roundhead prairieclover has short spikes that do not elongate in fruit. Also, the flowers of white prairieclover clearly exceed the broad bracts.

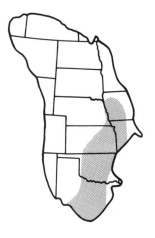

147

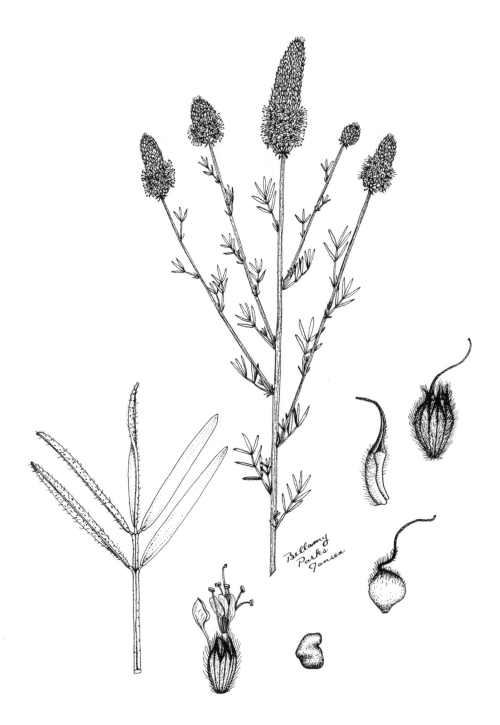

Figure 59   *Dalea purpurea*

Bellamy
Parks
Jansen

9. *Dalea purpurea* Vent.　　　　　　　　Purple prairieclover (Figure 59)

[*purpureus* (Lat.): for purple of various shades, including violet.]

Life Span: perennial. *Origin*: native. *Height*: 2–9 dm. *Stems*: few to many from a tough root system, eventually developing an underground caudex, herbaceous, erect (occasionally ascending or even prostrate), thinly pilosulous to glabrous, striate-ribbed, with scattered brownish-black glands. *Leaves*: alternate, numerous, 1–4 cm long, odd-pinnately compound; leaflets 3–7, usually 5, linear, (5)10–25(28) mm long, 0.5–1.5 mm wide, margins involute, midrib not visible on upper surface, upper surface usually glabrous, glabrous to sparingly hairy and glandular-punctate beneath; petiole similar to leaflets, involute. *Inflorescences*: terminal spikes, numerous, many-flowered, cylindric, 1–5(7) cm long, 7–14 mm wide; bracts 2–8 mm long, 1–2 mm wide, abruptly contracted into a recurved or erect tail, the dark brown tip densely pilosulous. *Flowers*: reddish-purple; calyx densely villous; petals 6 mm long, 4 of the petals and the 5 stamens joined near the tip of the calyx; banner separate. *Fruit*: legume, obliquely ovate, enclosed by the persistent calyx, 2–2.5 mm long, 1-seeded; seeds yellowish-green to brown, 1.5–2 mm long, punctate. 2n=14.

This species is sometimes divided into var. *purpurea* and var. *arenicola* (Wemple) Barneby. Var. *arenicola* has spikes 6.5–9 mm in diameter, and var. *purpurea* has spikes 9–12 mm in diameter. This difference probably does not warrant recognition.

Synonyms: *Petalostemon purpureus* (Vent.) Rydb., *P. mollis* Rydb.
Other Common Names: violet prairieclover, red tasselflower, wanahcha (Lakota)

Purple prairieclover is common on dry prairies throughout the Great Plains, and it blooms from June to early August (rarely until mid-September). It produces excellent forage for livestock and wildlife and may be an important component of upland prairie hay. It is high in protein and highly palatable, although it may cause bloat. It decreases with heavy grazing.

Seed is commercially available from numerous sources. Germination is enhanced by scarification and stratification. Purple prairieclover is used in seed mixtures for revegetation and reclamation. It is also used in perennial flower beds.

American Indians ate fresh and boiled purple prairieclover leaves. Bruised leaves were steeped in water and applied to fresh, open wounds. Comanche and Ponca chewed the roots for their pleasant flavor and made tea from the leaves. Pawnee used the stems to make brooms.

149

Figure 60   *Dalea villosa*

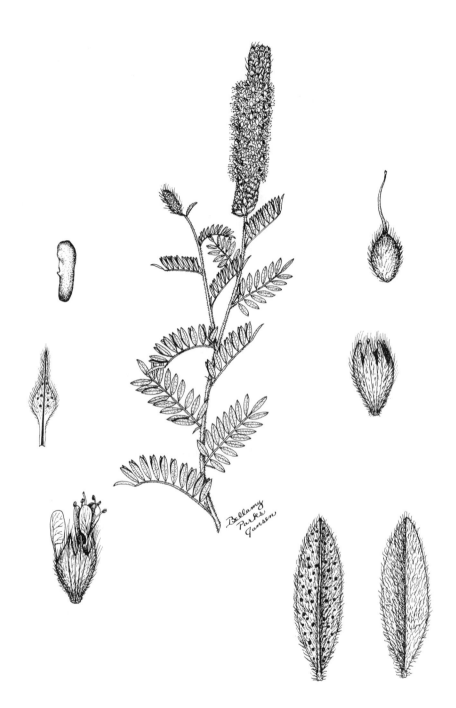

10. *Dalea villosa* (Nutt.) Spreng. Silky prairieclover (Figure 60)

[*villous* (Lat.): hairy or shaggy, referring to the densely villous covering on the plant.]

Life Span: perennial. *Origin*: native. *Height*: 1.5–4(6) dm. *Stems*: herbaceous, erect or ascending (rarely prostrate), densely villous, gray-green, simple below, arising from a woody caudex and taproot. *Leaves*: alternate, shortly petiolate, odd-pinnately compound, 1.5–4 cm long; leaflets 11–25, densely villous, gray-green, oblong to elliptic, usually acute, glandular-punctate beneath, usually flat (even when dry), 5–14 mm long, 1–4 mm wide. *Inflorescences*: spikes, terminating stems and branches, cylindric, 2–10 cm long, 7–9 mm wide, densely villous; bracts narrowly lanceolate, 1.5–5 mm long, surpassing the calyx, villous, caducous; peduncles up to 2.5 cm long. *Flowers*: pale rose to rose-purple; calyx 10-ribbed but the ribs obscured by pubescence, tube 2–3 mm long; lobes 5, 1–1.5 mm long; banner 4.5–5.5 mm long, claw 2–3 mm long; wings and keel 2.5–4.5 mm long; stamens 5, anthers orange, conspicuous. *Fruit*: legume, 2.5–3 mm long, densely villous, 1-seeded; seeds brown, smooth, narrowly ovoid, short beak, 2–2.5 mm long. $2n=14$.

Synonym: *Petalostemon villosus* Nutt.
Other Common Names: hairy prairieclover, casmu hoholhota (Lakota)

Silky prairieclover is common on sandy prairie, sandy woodlands, and on the fringes of blowouts. It flowers from July to August. Domestic livestock occasionally eat it. Deer, elk, and pronghorn graze it. Birds and rodents eat the seeds.

Commercial seed is infrequently available. It is promising for erosion control. Its gray-green color and attractive flowers make it potentially valuable for landscaping.

Lakota Indians consumed the roots for a laxative. They ate the leaves and flowers to reduce swelling inside of the throat.

Dalea tenuifolia (A. Gray) Shinners, slimflower prairieclover, is similar but has 3–11 leaflets per leaf, foliage is rarely villous-pilose, and the plants are not found in loose sand. Slimflower prairieclover grows in the southwestern Great Plains.

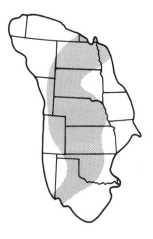

10. DESMODIUM Desv.

[*desmos* (Gk.): a bond or connection, referring to the jointed fruit.]

Perennial herbs, erect, prostrate, or trailing, usually with uncinate hairs; leaves usually alternate, pinnately 3-foliate; stipules ovate to subulate; subulate stipels usually present also; axils pubescent, partially with uncinate hairs; flowers in elongate panicled racemes; calyx tube more or less bilabiate, slightly oblique, upper 2 lobes connate for all or most of their length, lower 3 separate, the middle lobe often the longest; corolla papilionaceous, small, white to purple or violet, sometimes marked with yellow, drying to blue or yellow; banner oblong to nearly orbicular, narrowed at base; wings oblong; keel nearly straight; stamens usually diadelphous, sometimes monadelphous, median upper filament free; ovary short; fruits indehiscent, transversely segmented (loments), elevated on a stipe above the persistent calyx, eventually separating into 1-seeded joints, beset with uncinate hairs, sutures between segments usually more shallow on dorsal margin than on ventral margin.

More than 200 species occur in North and South America, Africa, Asia, and Australia. About 30 species are in North America. Thirteen are recorded in the Great Plains, but only 7 are relatively common.

A. Calyx lobes less than 1/2 as long as the tube; stipe of loment 3 times as long as calyx, longer than persistent remains of stamens; stamens monadelphous; dorsal margin of loment glabrous
    B. Flowering stem usually leafless, simple . . . . .5. *D. nudiflorum*
    B. Flowering stem leafy, branched. . . . . .3. *D. glutinosum*
A. Calyx lobes more than 1/2 as long as the tube; stipe of loment less than 2 times as long as calyx, shorter than remains of stamens; stamens diadelphous, dorsal margin of loment pubescent
    C. Petioles of leaves up to 3 mm long or less. . . . . .7. *D. sessilifolium*
    C. Petioles of leaves 3 mm long or longer
        D. Leaflets with uncinate hairs beneath
            E. Stem usually unbranched and with one inflorescence; leaflets reticulate-veined beneath . . . . .4. *D. illinoense*
            E. Stem usually much branched and with several inflorescences; leaflets not reticulate-veined beneath. . . . . .2. *D. canescens*
        D. Leaflets without uncinate hairs beneath
            F. Leaflets about 3 times as long as wide; lower margin of loment segments gradually curved, segments nearly semicircular . . . . .1. *D. canadense*
            F. Leaflets about 5 times as long as wide; lower margin of loment segments abruptly curved, segments more or less triangular . . . . . . .6. *D. paniculatum*

Figure 61   *Desmodium canadense*

1. *Desmodium canadense* (L.) DC.   Canada tickclover (Figure 61)

[*canadense*: of or from Canada.]

Life Span: perennial. *Origin*: native. *Height*: 4–20 dm. *Stems*: 1 to few from a caudex and taproot, herbaceous, erect, branched above, uncinulate-puberulent. *Leaves*: alternate, pinnately 3-foliate, 6–15 cm long (with petiole); leaflets lanceolate to lance-oblong, occasionally oblong, apex acute to slightly obtuse, base obtuse to broadly acute, uncinulate-puberulent and appressed-pilose on upper surface, pilose below, ciliate; terminal leaflet 5–10 cm long, 2–4 cm wide; petioles 2–30 mm long, longer below; petiole and leaf rachis together 1/6 to 1/2 as long as terminal leaflet; stipules linear-subulate, deciduous, 4–10 mm long, 1–1.5 mm wide at the base, ciliate, otherwise glabrous; stipels lanceolate, 2–4 mm long, persistent. *Inflorescences*: panicle of racemes, densely flowered, terminal; bracts conspicuous, ovate-lanceolate. *Flowers*: calyx tube campanulate, 1.5–2 mm long, bilabiate, upper lobe 4.5–5 mm long, middle tooth of lower lobe 5–7 mm long; corolla papilionaceous, 10–13(15) mm long, reddish-violet (rarely white) fading to purple or blue; stamens diadelphous. *Fruit*: loment; joints 1–5, each joint 5–7 mm long, 4–5 mm wide, slightly convex on one margin and quite obtuse on the other, uncinulate-pubescent; stipe 2–3 mm long; seeds brown, ovoid to broadly ellipsoid, smooth, 3–4 mm long. n=11.

Synonym: *Meibomia canadensis* (L.) O. Ktze. *Other Common Names*: Canada ticktrefoil, beggar's lice

Canada tickclover is infrequent to common in prairies and thickets and along rivers and roads. It is most common in sandy soil. It flowers from July to September. Immature plants are grazed by domestic livestock. It decreases with continued heavy grazing. Foliage is eaten by deer and rabbits. Many kinds of rodents and birds eat the seeds.

Canada tickclover is poor for erosion control. It has no value for landscaping.

Three other species have fruits similar to the above. *Desmodium marilandicum* (L.) DC., Maryland tickclover, has small (4–7 mm long), reddish flowers. It is similar to *D. ciliare* (Muhl. *ex* Willd.) DC., slender tickclover. It is difficult to separate the two, although *D. ciliare* has pink or white flowers. *Desmodium obtusum* (Muhl. *ex* Willd.) DC., obtuse tickclover, has small flowers (5–8 mm long) and relatively large terminal leaflets (3.5–7.5 cm long). All three species grow in the southeastern Great Plains.

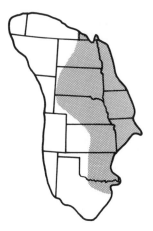

Figure 62   *Desmodium canescens*

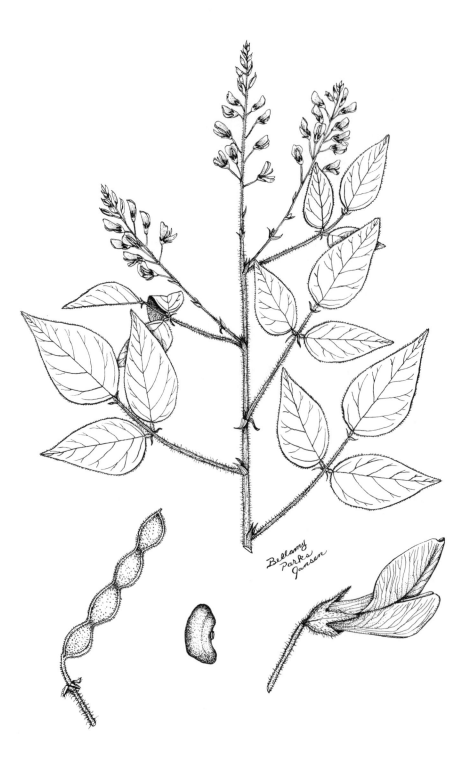

2. *Desmodium canescens* (L.) DC.

Hoary tickclover (Figure 62)

[*canescens* (Lat.): becoming gray, referring to the color caused by the pubescence.]

Life Span: perennial. **Origin**: native. **Height**: 5–15 dm. **Stems**: 1–10 from a caudex and taproot, herbaceous, erect, more or less pubescent. **Leaves**: alternate, pinnately 3-foliate, 6–16 cm long (with petiole); leaflets ovate to ovate-lanceolate, nearly smooth, terminal leaflet 3–12 cm long, 1.5–4.5(6) cm wide, usually 1/2 as wide as long, uncinulate-pubescent at least beneath, apex obtuse to acute, sometimes mucronate, rounded to cuneate at base; lateral leaflets inconspicuously asymmetrical; rachis stalk of terminal leaflet 1–3.5 cm long; petioles of principal leaves (2)5–10(12) cm long, nearly as long as the lateral leaflets; stipules long-persistent, ovate to ovate-lanceolate, acuminate, 5–12 mm long, 3–6 mm wide, ciliate, otherwise glabrous, base partially clasping; stipels persistent, linear, 2–7 mm long. **Inflorescences**: panicle of few to several racemes, racemes simple or rarely branched, axis villous or hirsute. **Flowers**: calyx tube campanulate, bilabiate, 1.5–2.5 mm long, sparsely hirsute to densely villous; lobes longer than the tube, 1.5–5 mm long; corolla papilionaceous, 8–12 mm long, pink to whitish-pink, drying to purple, wings and keel extending beyond the banner; stamens diadelphous; pedicels slender, 5–10(15) mm long, subtended by bracts; bracts ovate-lanceolate, 3–7 mm long, caducous. **Fruit**: loment uncinulate-pubescent; joints (2)4–6, semirhomboidal, 7–13 mm long, 4–5 mm wide; margins pubescent, upper margin convex to straight, more deeply indented on the lower side; stipe 1–3 mm long, seeds smooth, brown, reniform, 3.5–4.5 mm long, 2 mm wide. n=11.

Synonym: *Meibomia canescens* (L.) O. Ktze.

Other Common Names: tick trefoil, beggar's lice

Hoary tickclover is widely scattered in moist or dry soil of open woods and prairie. It is also found along streams, railroads, and roads. It flowers from July to September. Domestic livestock utilize the immature plants. It is generally not found where grazing is heavy. Deer have been reported to eat the leaves, while eastern turkey eat the flowers. Numerous rodents and birds consume the seeds.

Hoary tickclover has no value for landscaping. It is not effective for erosion control.

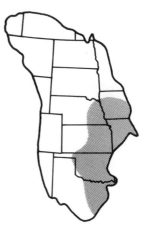

Figure 63   *Desmodium glutinosum*

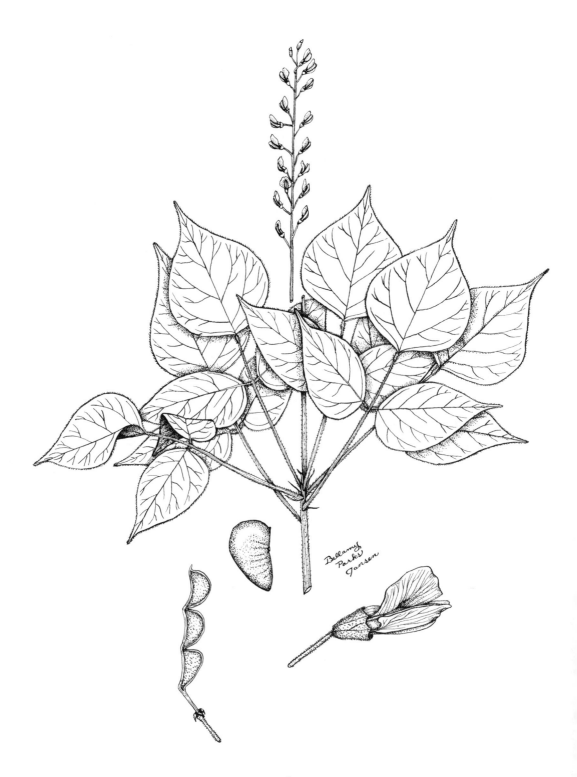

3. *Desmodium glutinosum* (Muhl. *ex* Willd.) Wood

Largeflower tickclover
(Figure 63)

[*glutinosus* (Lat.): sticky, referring to the hooked hairs covering the loments.]

Life Span: perennial. **Origin**: native. **Height**: (1)4–10 dm. **Stems**: herbaceous, erect from a caudex, unbranched, minutely uncinulate-puberulent. **Leaves**: alternate, pinnately 3-foliate, 1.5–3 dm long (with petiole), crowded into a single whorl at the top of the stem; leaflets sparsely pubescent; terminal leaflet round to ovate, 7–15 cm long, 5–12 cm wide, long acuminate; lateral leaflets slightly asymmetrical, ovate, acuminate; petioles 6–14 cm long; stipules linear, 8–13 mm long, semipersistent. **Inflorescences**: terminal raceme or panicle of racemes, 3–8 dm long, elevated above the leaves on a naked peduncle; bracts caducous, 5–10 mm long. **Flowers**: calyx tube 1.5–2.5 mm long, slightly irregular, lobes much shorter than the tube; corolla papilionaceous, 6–8 mm long, wings and keel slightly longer or equaling the banner, pink to purple (rarely white), stamens monadelphous. **Fruit**: loments, uncinulate-puberulent, seldom with more than 3 joints; joints semiobovate, 8–12 mm long, 4–6 mm wide; upper margins glabrous, straight or concave; lower margin U-shaped; stipe 5–10 mm long; seeds irregular, yellow to brown, flattened, 6–7 mm long, 4–5.5 mm wide. n=11.

Synonyms: *Desmodium acuminatum* (Michx.) DC., *Meibomia acuminata* (Michx.) Blake

Largeflower tickclover is common in rich woods and wooded valleys. It flowers from July to September. Domestic livestock graze the immature plants. They are also eaten by deer. Birds and rodents utilize the seed.

Seed is not commercially available.

Largeflower tickclover has no value for landscapes or erosion control.

Desmodium pauciflorum (Nutt.) DC., few-flowered tickclover, also has relatively small calyx lobes. It differs from *D. glutinosum* in having white flowers, puberulent loment stipes, scattered leaves, and terminal leaflets that are longer than wide. It grows in the eastern Great Plains.

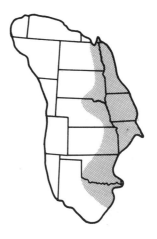

Figure 64   *Desmodium illinoense*

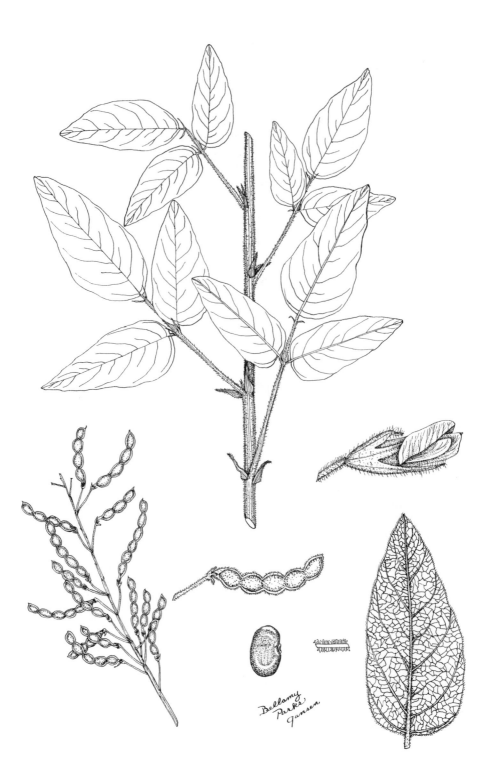

4. *Desmodium illinoense* A. Gray

Illinois tickclover (Figure 64)

[*illinoense*: of or from Illinois.]

Life Span: perennial. *Origin*: native. *Height*: 1–2 m. *Stems*: herbaceous, erect, stout, with uncinate hairs, unbranched from a caudex and taproot. *Leaves*: alternate, pinnately 3-foliate, 7–15 cm long (with petiole); leaflets coriaceous, ovate-lanceolate, apex acute to obtuse, often mucronate, with uncinate hairs on both sides, also with glandular hairs and strongly reticulate beneath; terminal leaflet (3)6–10 cm long, 1.5–6 cm wide; petioles nearly as long to longer than the terminal leaflet, pubescent with uncinate hairs; stipules ovate-acuminate, 9–15(20) mm long, 2–6 mm wide, semipersistent, ciliate, glabrous above, pubescent beneath. *Inflorescences*: often a single terminal, elongate raceme, sometimes a simply-branched panicle of a few racemes; axis villous to hirsute; bracts lanceolate, 1 cm long, caducous. *Flowers*: calyx persistent, bilabiate, tube 1–2 mm long, upper lobe 2–4.5 mm long, lower lobe 3–6 mm long; corolla papilionaceous, 8–10 mm long, pink to white, fading to purple; stamens diadelphous. *Fruit*: loments with 3–7 joints, each joint 4.5–8 mm long, rounded on both margins; margins uncinulate-pubescent; stipe less than 1 mm long; seeds dark-brown, plump, ovoid to reniform, smooth, 3–3.5 mm long, 2 mm wide. n=11.

Synonym: *Meibomia illinoense* (A. Gray) O. Ktze.
Other Common Name: tick trefoil

Illinois tickclover is scattered to locally common in rich prairie soils of ravines and hillsides. It is not common in waste areas and woods or on roadsides. It flowers from June to September. Domestic livestock and deer graze the immature plants. It rapidly decreases in abundance with grazing. Birds and rodents eat the seeds, and it is a good honey plant.

Seed is not commercially available. Illinois tickclover has no value for landscaping or erosion control.

Figure 65   *Desmodium nudiflorum*

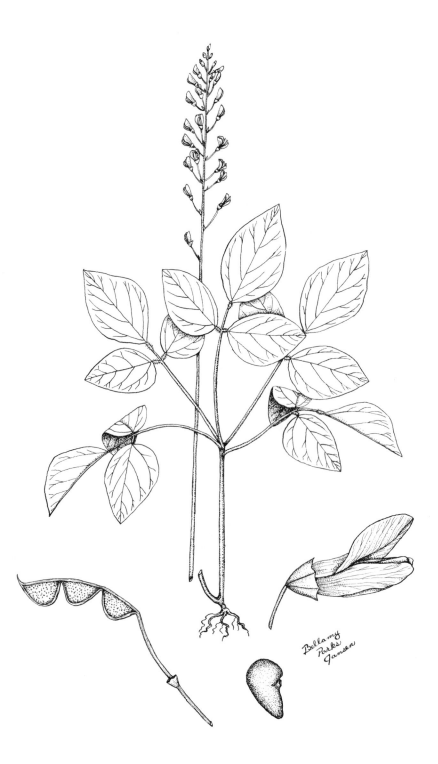

5. *Desmodium nudiflorum* (L.) DC. Barestem tickclover (Figure 65)

[*nudus* (Lat.): naked; + *florus* (Lat.): flower; in reference to the sterile branch.]

Life Span: perennial. **Origin**: native. **Height**: 4–10 dm. **Stems**: herbaceous, erect or ascending, forked from the base; one branch sterile, 1–3 dm tall, naked below, bearing a crowded cluster of leaves at the summit; fertile branches 1–4, scapose, ascending, about 3 times the height of the sterile stems. **Leaves**: alternate, pinnately 3-foliate; lateral leaflets ovate-oblong to ovate; terminal leaflet elliptic to ovate, 4–10 cm long, obtuse, short acuminate or acute; petioles 4–12 cm long; stipules caducous. **Inflorescences**: usually a raceme, occasionally a panicle of few racemes, leafless; bracts linear-lanceolate, 4–6 mm long, caducous. **Flowers**: calyx tube 1.5–2.5 mm long, lobes hardly evident; corolla papilionaceous, 6–9 mm long, petals nearly equal in length, pink to purple (rarely white); stamens monadelphous; pedicels 1–2 cm long, capillary. **Fruit**: loments with 2–4 joints, each joint 8–11 mm long, 4–5 mm wide, semiobovate; margins glabrous, upper margin nearly straight; surfaces uncinulate-puberulent; stipe glabrous, 8–12 mm long; seeds variable, yellow to brown, flat, smooth, 5–6 mm long. n=11.

Synonym: *Meibomia nudiflora* (L.) O. Ktze.
Other Common Name: scapose tickclover

Barestem tickclover is not common in the Great Plains. It is found in rich soils of woods and wooded slopes. It flowers from July to September. Domestic livestock will eat the plants, but they generally do not grow in pastures or prairies. Deer eat the foliage, and numerous birds and rodents consume the seeds.

Barestem tickclover seed is not commercially available. It has no value for landscaping or erosion control.

Figure 66 *Desmodium paniculatum*

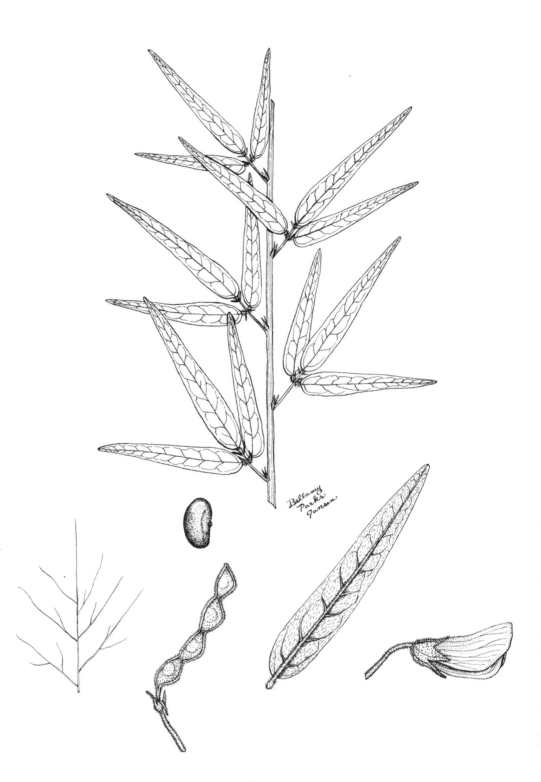

6. *Desmodium paniculatum* (L.) DC.

Panicled tickclover (Figure 66)

[*paniculus* (Lat.): tuft, referring to having panicles.]

Life Span: perennial. **Origin**: native. **Height**: 5–12(14) dm. **Stems**: 1 to several from a caudex and taproot, herbaceous, slender, erect, ridged, usually branched above, uncinulate-puberulent to pilose. **Leaves**: alternate, pinnately 3-foliate, 2.5–15 cm long (with petiole); leaflets variable, thin, lanceolate or oblong-lanceolate to linear; terminal leaflet 2–10 cm long, 1–4 cm wide, obtuse or nearly acute, sparsely hairy to tomentose on the lower surface, not uncinulate-puberulent, lateral veins arcuate-ascending; petioles of principal leaves 2–6 cm long; stipules lanceolate-subulate, 2–4(6) mm long, caducous to semipersistent. **Inflorescences**: terminal panicle, 1–4 dm long, much-branched. **Flowers**: calyx tube 1.5–2 mm long, bilabiate, upper lobe 1.5–3.5 mm long, lower lobe 2.5–5.5 mm long; corolla papilionaceous, 6–8 mm long, lavender to reddish-purple (rarely white), fading to blue; stamens diadelphous; pedicels 4–11 mm long. **Fruit**: loments, surface and margins uncinulate-puberulent with 3–6 joints, each joint 5–10 mm long, 3.5–4.5 mm wide, slightly angled or convex on one margin, angled on the second margin, more or less triangular; stipe 2.5–3.5 mm long, exceeding the persistent calyx; seeds reddish-brown or tan, 3.5–4.5 mm long, reniform to elliptic. n=11.

Plants of this species are highly variable, especially in pubescence and leaflet shape. Some taxonomists have divided this entity into 4 species. McGregor, in the *Flora of the Great Plains*, separated them into 2 varieties: var. *dillenii* (Darl.) Isley, with terminal leaflets generally 1.5–2.5(3) times longer than wide, and var. *paniculatum*, with terminal leaflets 3–8 times longer than wide.

Synonyms: *Meibomia paniculata* (L.) O. Ktze., *M. pubens* (T & G.) Rydb., *M. dillenii* (Darl.) O. Ktze., *Desmodium dillenii* Darl., *D. perplexum* Schub., *D. glabellum* (Michx.) DC.

Panicled tickclover is infrequent to locally common in dry woods and prairie, especially if the soil is rocky or sandy. It occasionally is found on roadsides. It flowers from July to September. It is consumed by domestic livestock and deer while it is immature. Rodents and birds utilize the seeds.

Seed is not commercially available. Panicled tickclover has no value for landscaping or erosion control.

Desmodium cuspidatum (Muhl. *ex* Willd.) Loud., longleaf tickclover, also has leaflets with uncinate hairs beneath, but it has long and persistent stipules (8–18 mm long) and pink flowers.

Figure 67   *Desmodium sessilifolium*

7. *Desmodium sessilifolium* (Torr.) T. & G. Sessile tickclover (Figure 67)

[*sessilis* (Lat.): pertaining to sitting; + *folium* (Lat.): a leaf; referring to the close attachment of the leaf blade to the stem.]

Life Span: perennial. **Origin**: native. **Height**: 8–15 dm. **Stems**: 1 to several from a caudex and taproot, herbaceous, usually simple to the inflorescence, densely uncinulate-puberulent. **Leaves**: alternate, pinnately 3-foliate, nearly sessile; leaflets oblong to oblong-lanceolate, apex acute to obtuse, usually mucronate, uncinulate-puberulent, sparsely pubescent or sometimes glabrous above, reticulate-veined beneath; terminal leaflet 1.5–9 cm long, 5–18 mm wide; lateral leaflets 3–6 cm long; petioles 1–5 mm long, shorter than stalk of terminal leaflet; stipules lanceolate, 3–5(10) mm long, semipersistent to caducous. **Inflorescences**: panicle of racemes, 1–3 dm long, branches ascending; pedicels 1–5 mm long. **Flowers**: calyx tube 1–1.5 mm long, bilabiate; lobes often longer than the tube, upper lobe entire or notched; corolla papilionaceous, 4–6 mm long; petals pink, lavender, or white; stamens diadelphous. **Fruit**: loments with 1–3(4) joints; each joint 4–6.5 mm long, 3.5–4.5 mm wide, rounded above and below; margins and surface uncinulate-puberulent; stipe 1–2 mm long; seeds light-tan to olive or brown, narrowly ovoid, smooth, slightly flattened, 2.5–3.5 mm long. n=11.

Synonym: *Meibomia sessilifolia* (Torr.) O. Ktze.
Other Common Name: sessileleaved tickclover

Sessile tickclover is scattered to locally common in dry or sterile soil of open woodlands, prairie hillsides, ravines, valleys, and roadsides. It flowers from July to September. It is eaten by domestic livestock and deer when it is immature. It rapidly disappears with continued grazing. Rodents and birds utilize the seed. It is a fair to good honey plant.

Sessile tickclover seed is not commercially available. It has no value for erosion control or landscaping.

Desmodium rotundifolium DC., dollarleaf, is similar, but it has prostrate stems forming mats 1.5–2.5 m in diameter, and flowers are usually purple. It grows in the extreme eastern Great Plains.

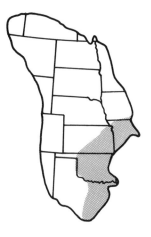

167

11. GLYCYRRHIZA L.

[*glykys* (Gk.): sweet; + *rhiza* (Gk.): root.]

Glandular-dotted perennial herbs; leaves odd-pinnately compound; inflorescences of dense axillary racemes; calyx 5-lobed, bilabiate, upper 2 lobes fused part of their length, tube campanulate; petals usually acute; banner oblong-obovate, tapering to the base; wings and keel shorter, oblong, clawed at base; stamens diadelphous for 1/2 their length; anthers alternately large and small; ovary cylindric; legumes tardily dehiscent to indehiscent, somewhat inflated, slightly flattened, glandular or spiny, few-seeded.

About 15 species have been described, and they have wide distribution. Only 1 species occurs in the Great Plains. *Glycyrrhiza glabra* L. is the licorice of commerce. It is grown commercially in North America.

Figure 68   *Glycyrrnhiza lepidota*

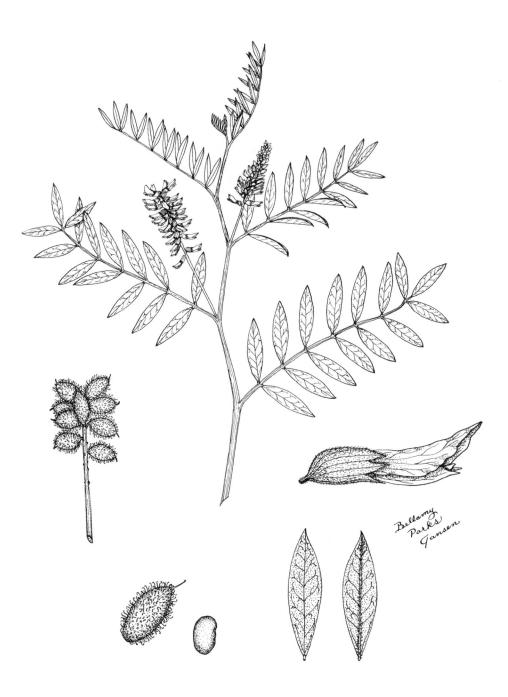

Bellamy
Parks
Jansen

Glycyrrhiza lepidota Pursh

Wild licorice (Figure 68)

[*lepidotos* (Gk.): scaly, referring to the leaves being minutely scaly when young.]

Life Span: perennial. **Origin**: native. **Height**: 3–10(12) dm. **Stems**: 1 to several from long rhizomes, herbaceous, little branched. **Leaves**: alternate, odd-pinnately compound; leaflets (7)11–19(21), oblong-lanceolate, 2–7 cm long, 4–20 mm wide, dotted with minute glands, mucronate, midrib often pubescent on lower side; petioles 0.5–5 cm long; stipules 3–7 mm long, slender, deciduous. **Inflorescences**: axillary racemes, erect, dense, 2.5–5 cm long, shorter than subtending leaves, peduncled. **Flowers**: calyx tube campanulate, 5–7 mm long, lobes 2.5–3 mm long; petals yellowish-white, acute; banner oblong-obovate, tapering to the base, 9–14 mm long; wings and keel shorter (8–12 mm long), oblong, clawed at the base; stamens 10, diadelphous for 1/2 their length; anthers alternately large and small. **Fruit**: legume, oblong-ellipsoid, 1–2 cm long, reddish-brown at maturity, densely covered with hooked prickles, 3- to 5-seeded, indehiscent to tardily dehiscent; seeds olive to grayish-brown, short-reniform, plump, smooth, 2.5–4 mm long. n=8.

Var. *glutinosa* (Nutt.) S. Wats. is generally found west of the Great Plains. It has stalked glands on the petioles, upper stem, and inflorescences. Var. *lepidota* is the variety common to the Great Plains.

Other Common Names: winawizi cik'ala (Lakota), pithahatusakitstsuhast (Pawnee)

Wild licorice is common on moist prairies, rich shores, meadows, along railroads, and in waste places. The hooked prickles on the legume stick to animals assuring wide seed distribution. It flowers from May to early August. It has low palatability to domestic livestock, although it may be consumed in dry hay. Deer and pronghorn consume the foliage. Birds and rodents eat the seeds, and the roots are eaten by plains pocket gophers.

Seeds are generally not commercially available. Germination is low and can be improved by scarification. It has limited applications in landscaping. It may be started from rhizome cuttings in fall or spring, and stem cuttings are relatively easy to start in a mist bench.

Wild licorice was widely used as medicine by Indians on the Great Plains. A poultice was applied to open wounds to stop bleeding and to horses' backs to relieve soreness. Lakota used a decoction as a fever remedy for children. Steeped leaves were applied to ears to relieve earache. Roots were chewed and held in the mouth to relieve toothache and sore throats. Roots were also eaten either raw or baked for nourishment.

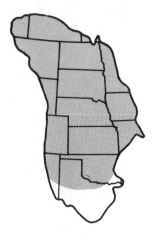

12. LATHYRUS L.

[*lathyros* (Gk.): ancient name of some leguminous plant.]

Annual or perennial herbs; stems decumbent, erect, or ascending, usually twining, sometimes winged; leaves alternate, even-pinnately compound, with 2–18 leaflets, usually terminated by a tendril or sometimes a bristle; stipules hastate to obliquely semi-sagittately lobed, persistent; inflorescence an axillary raceme of few to many flowers; calyx irregular or regular, tube campanulate to somewhat turbinate, lobes equal or upper 2 shorter; corolla papilionaceous, reddish-purple to white or yellow; banner broadly obovate to rotund; wings narrowly to broadly obovate; keel upwardly curved; stamens 10, diadelphous; ovary sessile or stipitate; style twisted or straight, bearded along inner side; legumes terete to flat, thin-walled, dehiscent, 2- to many-seeded.

A genus of nearly 200 species found primarily in the northern temperate zone and in South America. Of the 6 species found in the Great Plains, 3 are considered to be common.

A. Leaflets on mature leaves only 2; calyx regular; stem winged . . . . .1. *L. latifolius*
A. Leaflets on mature leaves 4 or more; calyx regular or irregular; stems angled, not winged
    B. Tendrils branched . . . . .3. *L. venosus*
    B. Tendrils bristle-like, simple . . . . .2. *L. polymorphus*

Figure 69   *Lathyrus latifolius*

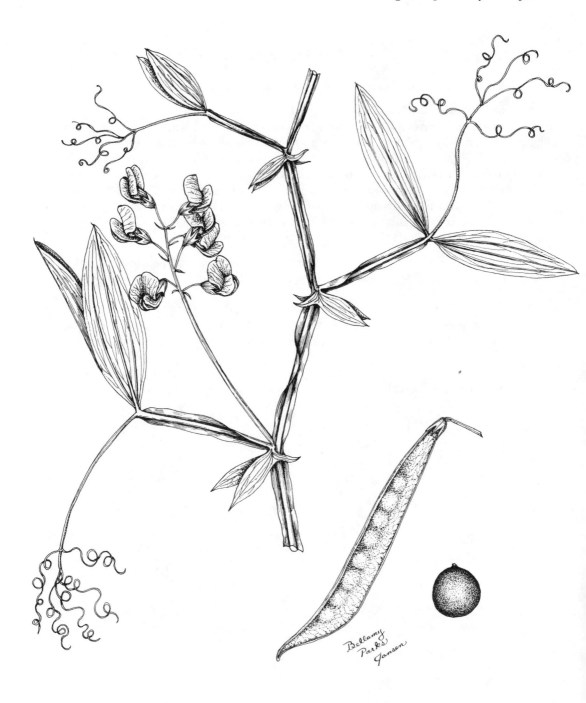

1. *Lathyrus latifolius* L.

Perennial sweetpea (Figure 69)

[*latus* (Lat.): broad; + *folium* (Lat.): leaf, in reference to the broad leaves.]

Life Span: perennial. *Origin*: introduced (from southern Europe). *Height*: 5–20 dm. *Stems*: herbaceous, arising from rhizomes, climbing or trailing, glabrous, slightly glaucous, broadly winged, wings 5–10 mm wide. *Leaves*: alternate, pinnately compound, 2-foliate, glabrous; leaflets lanceolate to elliptic, 4–8 (15) cm long, 1–3(5) cm wide, mucronate, with prominent parallel veins; tendrils much branched; petiole 3–9 cm long, broadly winged, about as wide as the stem; stipules lanceolate with a basal lobe, semisagittate, foliaceous, conspicuously veined, 1.5–4(8) cm long, usually wider than the stem. *Inflorescences*: axillary racemes, with 4–10(14) flowers; peduncles 1–2 dm long. *Flowers*: calyx tube campanulate, 4–6 mm long, persistent; lobes unequal, upper lobes 2–4 mm long, lateral and lower lobes 4–10 mm long; petals purple (less commonly red, white, or pink); banner strongly reflexed, 1.5–2.5 cm long, nearly as wide as long, clawed; wings and keel clawed; pedicels 8–15 mm long; bracts setaceous to linear-subulate, 2–6 mm long. *Fruit*: glabrous legume, 6–10 cm long, 7–10 mm wide, reticulately nerved, flattened, 6- to 25-seeded; seeds dark brown, rugose, spherical or oblong, 5 mm long. n=7.

Other Common Names: everlasting pea, perennial peavine

Perennial sweetpea has occasionally escaped from cultivation. It is infrequent along fence rows and roadsides. It is also found in waste places near homes or old home sites. It flowers from May to September. Domestic livestock and wildlife eat the herbage, but the seeds have been reported to be highly toxic. Neurolathyrism is an imbalance in the nervous system, and osteolathyrism causes skeletal deformities and aortic rupture. Horses are especially susceptible.

Perennial sweetpea is commonly grown in gardens. Several varieties are commercially available. It has also been seeded on highway embankments for erosion control. Germination is improved by scarification.

Lathyrus pusillus Ell., singletary vetchling, also has 2 leaflets per leaf. It is an annual with wingless petioles and small flowers (less than 1.5 cm long). It grows in the southeastern portion of the Great Plains.

175

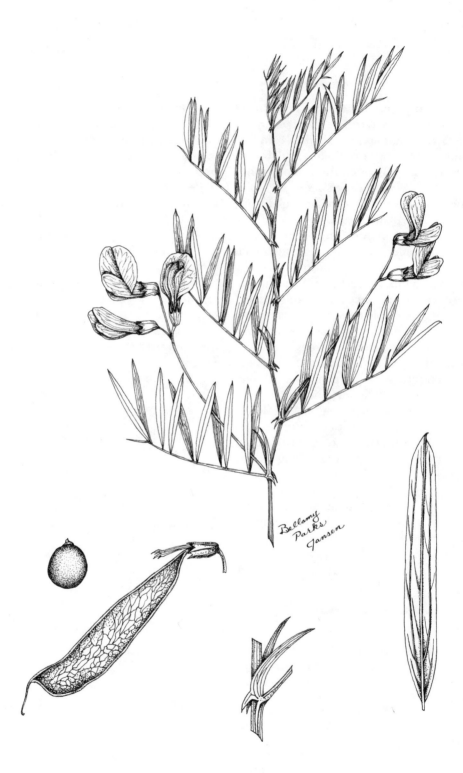

Figure 70   *Lathyrus polymorphus*

2. *Lathyrus polymorphus* Nutt.

Hoary vetchling (Figure 70)

[*poly* (Gk.): many; + *morphe* (Gk.): shape, in reference to the many shapes of the leaflets.]

Life Span: perennial. *Origin*: native. *Height*: 1–5 dm. *Stems*: herbaceous, glabrous to pubescent, erect to ascending from rhizomes and a branching caudex. *Leaves*: alternate, even-pinnately compound, glabrous (rarely pubescent); leaflets 4–12, scattered or paired, 1.5–5 cm long, 1–5 mm wide, linear-lanceolate to linear-elliptic, prominently veined; rachis extended to a bristle, lacking tendrils; stipules 7–21 mm long, lower lobe lanceolate or triangular, upper lobe lanceolate. *Inflorescences*: axillary racemes of 2–8 fragrant flowers; peduncles 6–7 cm long, usually surpassing the leaves. *Flowers*: calyx persistent, 4–6 mm long, upper lobes 2–3 mm long, lower lobe 4–6 mm long, lateral lobes 3–4 mm long; banner purple, 2–3 cm long, clawless; wings purple or blue to white; wings and keel 1.5–2.5 cm long. *Fruit*: legumes 2–6 cm long, 5–10 mm wide, coriaceous, few-seeded; seeds dark green to brown, nearly spherical, 5–6 mm long, smooth. n=7.

Two subspecies are recognized. Subsp. *polymorphus* is glabrous and is found in the eastern Great Plains. Subsp. *incanus* (Sm. & Rydb.) C. L. Hitchc. is pubescent and primarily grows in the western Great Plains.

Synonyms: *Lathyrus decaphyllus* Pursh, *L. hapemanii* A. Nels., *L. incanus* (Sm. & Rydb.) Rydb., *L. stipulaceous* (Pursh) Butt. & St. John

Hoary vetchling is locally common on dry, sandy prairies and in rocky, open woods. It is also found in stream valleys and on sand dunes. It flowers from May to June. It is good forage for cattle and sheep. The seeds are reported to be poisonous to horses, producing lameness.

Hoary vetchling seed is not commercially available. It has the potential to prevent erosion. It has little value for landscaping.

Lathyrus ochroleucous Hook., yellow vetchling, also has 4 or more leaflets. It differs from *L. polymorhpus* by having branched tendrils and white or ochroleucous flowers. It grows in the northern portion of the Great Plains.

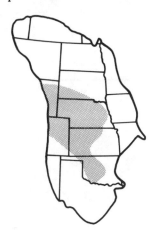

Figure 71   *Lathyrus venosus*

3. *Lathyrus venosus* Muhl. *ex.* Willd.   Bushy vetchling (Figure 71)

[*venosus* (Lat.): conspicuously veined, in reference to the petals.]

Life Span: perennial. *Origin*: native. *Height*: 0.5–1(2) m long. *Stems*: herbaceous, coarse, sprawling (sometimes erect), angled but not winged, branched, pubescent, arising from rhizomes. *Leaves*: alternate, even-pinnately compound; leaflets 8–14, scattered or in pairs, narrowly to broadly elliptic, 3–6 cm long, 1–3 cm wide, rounded to the mucronate summit; lower leaflet surface finely pubescent, lighter color; tendrils sparingly branched; petioles 0.5–1.5(3) mm long; stipules narrowly lanceolate to semiovate, 5–20(35) mm long. *Inflorescences*: racemes of (5)10–30 flowers, dense or loose; bracts minute, caducous; peduncles stout, usually shorter or equaling the subtending leaves. *Flowers*: purple; calyx tube campanulate, glabrous to shortly pubescent, 3.5–4.5 mm long, lobes unequal, lower lobes longer than the upper but shorter than the tube; banner 1.5–2 cm long, obcordate, claw nearly equaling the blade; wings shorter than the banner; keel shorter than the wings; pedicels 3–7 mm long. *Fruit*: legume, linear-oblong, 4–6 cm long, 5–8 mm wide, glabrous to short pubescent; seed dark-brown, smooth, 4–5 mm long, 2–3 mm wide. n=7.

The variety common to the Great Plains is *intonsus* Butt. & St. John.

Synonym: *Lathyrus oreophyllus* Woot. & Standl.
Other Common Names: wild peavine, viney peavine

Bushy vetchling is infrequent to locally common in prairie ravines, stream valleys, open woodlands, roadsides, and lake shores. It flowers from May to July. It is grazed by all classes of livestock. Deer eat the foliage, and plains pocket gophers consume the roots. It is important nesting cover for ring-necked pheasants and other upland birds.

Bushy vetchling seed is not available commercially. It has limited applications in erosion control and for landscaping.

Lathyrus palustris L., marsh vetchling, is similar to *L. venosus*. *Lathyrus palustris* usually has 8 or fewer leaflets and each raceme has 9 or fewer flowers. *Lathyrus venosus* has more than 8 leaflets, and each raceme usually has 10 or more flowers. *Lathyrus palustris* grows in the northeastern Great Plains.

13. LESPEDEZA Michx.

[*Lespedeza*: derived from a misspelling of Cespedes after V. M. de Cepsedes, Spanish governor of Florida, who aided Andre Michaux's 1785–96 botanical exploration of North America.]

Annual or perennial herbs with woody rhizomes (some introduced shrubs), erect or ascending, often branched, pubescent or glabrous; leaves alternate, numerous, pinnately 3-foliate, stipulate but without stipels; leaflets entire, often mucronate; inflorescences of spicate or capitate racemes or of solitary flowers; flowers purple to yellowish-white; calyx tube campanulate, short, 5-lobed; lobes linear-subulate; banner suborbicular to oblong-obovate, spreading to erect, short-clawed; wings oblong, straight, clawed, connivent with the keel; keel petals obliquely obovate; stamens diadelphous; ovary short; legume oval to elliptic, indehiscent, 1-seeded, persistent; flowers of perennial species of 2 types, chasmogamous and cleistogamous; in white-flowered species, the chasmogamous, petalous flowers are fertile and most abundant; in purple-flowered species, the cleistogamous, apetalous flowers are most abundant and produce more mature fruits.

Most of the 140 species are native to Asia. Only 20 species are native to North America, and most grow in the eastern portion of the continent. Eleven species grow in the Great Plains, and 6 are most common.

A. Stipules brown, ovate-lanceolate, scarious, many-nerved, persistent, usually longer than the petioles; plants annual
   B. Petioles of the main stem leaves over 4 (up to 10) mm long; stems antrorsely appressed-pubescent . . . . .3. *L. stipulacea*
   B. Petioles of the main stem leaves 3 mm or less in length; stems retrorsely appressed-pubescent . . . . .4. *L. striata*
A. Stipules setaceous to subulate, soon withering or caducous; plants perennial
   C. Flowers cream to white (may be marked with purple or pink); calyx equaling or exceeding the mature legume
      D. Flowers borne singly or in axillary clusters of 2–3(4); leaflets 1–2.5 cm long; wings and keel equal . . . . .2. *L. cuneata*
      D. Flowers numerous in subglobose to short-ovoid heads; leaflets 2–4.5 cm long; wings longer than the keel . . . . .1. *L. capitata*
   C. Flowers purple; calyx 1/2 as long as mature legume or shorter
      E. Leaflets 3 or more times as long as wide; peduncles usually shorter than the subtending leaves . . . . .6. *L. virginica*
      E. Leaflets less than 3 times as long as wide; peduncles usually longer than the subtending leaves . . . . .5. *L. violacea*

Figure 72   *Lespedeza capitata*

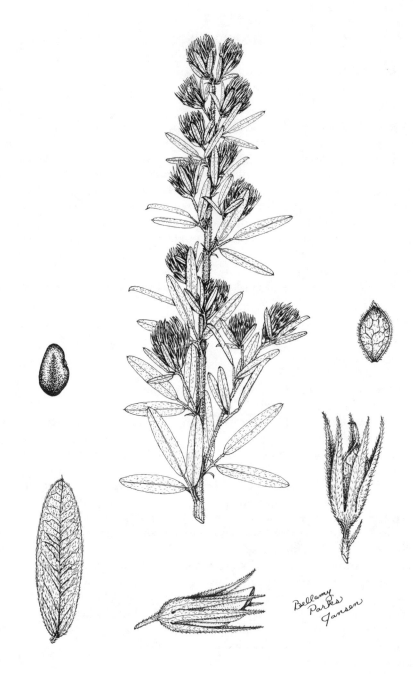

1. *Lespedeza capitata* Michx.   Roundhead lespedeza (Figure 72)

[*capitatus* (Lat.): headlike or globose, referring to the flower clusters.]

Life Span: perennial. **Origin**: native. **Height**: 6–15(20) dm. **Stems**: herbaceous, usually erect from a shortly branching caudex, rigid, simple or branched above, sparsely to densely silvery-pubescent, densely leafy. **Leaves**: alternate, pinnately 3-foliate, subsessile; leaflets elliptic to oblong-lanceolate, 2–4.5 cm long, 5–18 mm wide, obtuse and mucronate at the apex, nearly glabrous above, densely silvery-pubescent beneath; petioles 2–5 mm long, shorter than stalk of terminal leaflet; stipules filiform to subulate, 3–8 mm long, persistent. **Inflorescences**: sub-globose to short-ovoid racemes, 12–25 mm long, numerous, crowded; peduncles rarely longer than the racemes, usually shorter than the subtending leaves. **Flowers**: mostly chasmogamous; calyx tube 0.5–1 mm long, persistent, reddish-brown; lobes 5, 6–10(14) mm long, villous; corolla cream to white, 8–12 mm long; banner longer than the keel, sometimes with a purple spot, about as long as the calyx; wings longer than the keel; pedicels 1–3 mm long. **Fruit**: legume, elliptic to oblong, densely pubescent, shorter than the calyx, 4–7 mm long, 1-seeded; cleisto-gamous legumes scattered among the others, 4–5 mm long; seeds tan or green to brown or black, shiny, smooth, ellipsoid, slightly flattened, 2.5–3 mm long. n=10.

Other Common Names: parus-as (Pawnee), tehutohi nuga (Omaha-Ponca)

Roundhead lespedeza is common on prairies, sand dunes, dry fields, sandy woods, and roadsides. It flowers from June to September. It is excellent forage for all classes of livestock. The foliage is eaten by deer and turkey. Seeds are consumed by upland birds and rodents.

Roundhead lespedeza seed is commercially available. It should be scarified to assure high rates of germination. It is important for soil stabilization. Roundhead lespedeza is seldom used for landscaping, but it is commonly used in dry arrangements because of its persistent, reddish-brown calyx.

Commanche made a beverage from the leaves. Omaha-Ponca burned pieces of stem into their flesh as counterirritants for rheumatism and neuralgia.

Two additional species are similar, *Lespedeza hirta* (L.) Hornem., hairy lespedeza, has terminal leaflets more than 1/2 as wide as long. It is rare in the eastern Great Plains. *Lespedeza leptostachya* Engelm., slenderspire lespedeza, has the stalk of the terminal leaflet which is shorter than the petiole. Its inflorescence is open and interrupted. It grows in the northeastern Great Plains and is endangered.

Figure 73 *Lespedeza cuneata*

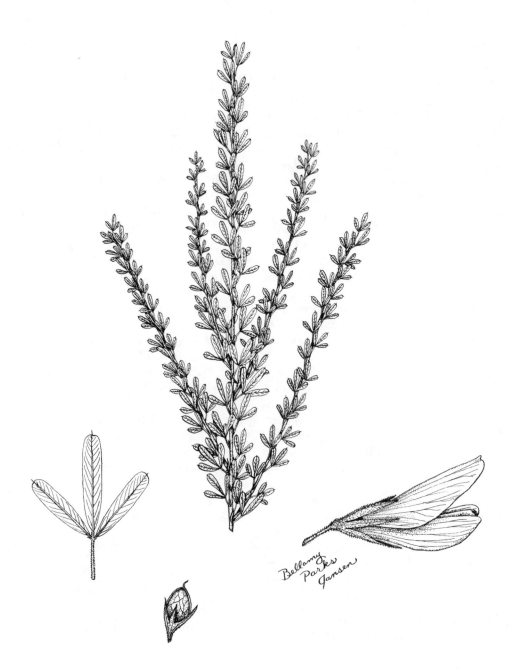

2. *Lespedeza cuneata* (Dumont) G. Don

Sericea lespedeza (Figure 73)

[*cuneata* (Lat.): wedge-shaped, in reference to the shape of the leaflets.]

Life Span: perennial. *Origin*: introduced (from eastern China, Korea, and Japan). *Height*: 5–20 dm. *Stems*: herbaceous to somewhat shrubby, erect, from a knobby caudex, with numerous elongate virgate branches, sulcate, pubescent on the angles. *Leaves*: alternate, pinnately 3-foliate; leaflets erect or ascending, linear-cuneate, 1–2.5 cm long, truncate and mucronate at the apex, glabrous above, sericeous beneath; petioles (0)2–5 mm long; stipules setaceous, 3–12 mm long. *Inflorescences*: axillary, mostly solitary or in clusters of 2–4. *Flowers*: chasmogamous flowers white or cream, marked with purple or pink along the veins of the banner; calyx tube 0.5–1 mm long, sericeous; lobes 5, lance-subulate, 3–5 mm long; banner 6–9 mm long; wings and keel equal, shorter than the banner; cleistogamous flowers common, scattered among the others. *Fruit*: oval legume, 2.5–3.5 mm long, glabrate or appressed-pubescent, 1-seeded; seeds brown to olivaceous, often mottled with brown, 1.5–2.5 mm long, ellipsoid to ovoid, slightly flattened. n=19.

Other Common Name: Chinese bushclover

Sericea lespedeza has escaped from seedings to roadsides, stream valleys, open woods, thickets, and waste places. It grows most frequently on well-drained clay loam or silt loam soils. It can grow on infertile soils with a low pH. It flowers from July to October. It produces forage of low quality. It contains condensed tannins that cause reduced digestibility by inhibiting cellulolytic enzymes. Low-tannin cultivars produce forage of acceptable quality for domestic livestock. Foliage is eaten by deer, rabbits, and eastern turkeys. Seeds are eaten by birds and rodents. It is a good honey plant.

Sericea lespedeza was first tested in North Carolina in 1896. Stem thickness is generally reduced in the several cultivars that have been developed. Seed should be scarified and inoculated. Establishment is slow. In the eastern states it is used to stabilize gullies and control erosion on highway embankments. It has little potential for landscaping.

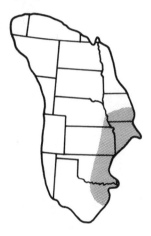

Figure 74   *Lespedeza stipulacea*

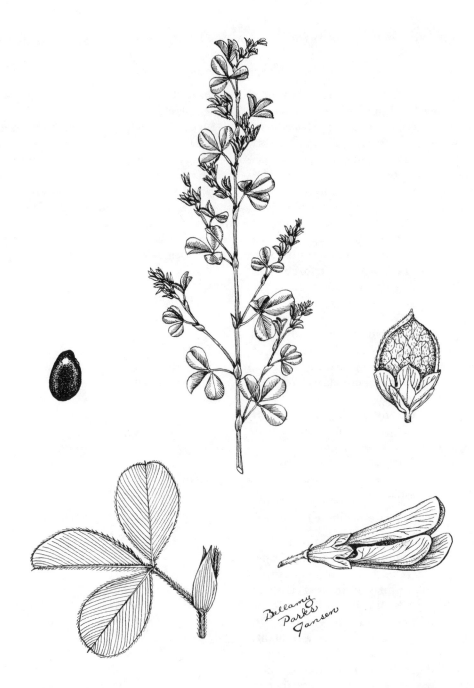

3. *Lespedeza stipulacea* Maxim.

[*stipula* (Lat.): small stalk, referring to the leaflike structures occurring at the base of the petioles.]

Life Span: annual. *Origin*: introduced (from eastern Asia, Korea in particular). *Height*: 1–6 dm. *Stems*: herbaceous, erect or ascending from taproots, diffusely branched from the base, sparsely pubescent with antrorsely appressed hairs. *Leaves*: alternate, pinnately 3-foliate, often appearing subpalmate; leaflets broadly obovate (6)10–15(25) mm long, 4–10 mm wide, apex obtuse to emarginate, glabrous or glabrate, ciliate on the margins and lower midvein; petioles 4–10 mm long; stipules ovate-lanceolate, brown, scarious, many-nerved, persistent, 4–6 mm long. *Inflorescences*: 1–3 flowers in upper axils, dense, leafy, chasmogamous and cleistogamous flowers mixed. *Flowers*: calyx tube 1 mm long; lobes 5, ovate, shorter than or about equaling the tube, upper lobe nearly completely connate; corolla 6–8 mm long, pink or purple. *Fruit*: legumes oval or obovate, 2.5–3 mm long, strongly reticulate, minutely appressed-pubescent, 1-seeded; seeds black or brown, shiny, ellipsoid, slightly flattened, 1.5–2 mm long. n=11.

Other Common Name: Korean clover

Korean lespedeza has escaped from seedings to disturbed prairies, open woods, roadsides, and waste ground. It flowers from July to October. It was introduced to North America in 1919 for pasture and soil improvement. It is occasionally cut for hay. It is palatable to all classes of livestock. Foliage is also eaten by deer and rabbits. Seeds are eaten by birds and rodents.

Korean lespedeza does not cause bloat. Several instances of hemorrhagic syndrome have been reported. The condition is similar to sweetclover poisoning and is caused by moldy hay.

Korean lespedeza is a valuable plant for soil conservation. Several cultivars are available. Seed should be scarified to improve germination. It has essentially no value for landscaping.

Figure 75   *Lespedeza striata*

4. *Lespedeza striata* (Thunb.) H. & A.   Common lespedeza (Figure 75)

[*stria* (Lat.): furrowed or marked with fine longitudinal lines, in reference to the stipules.]

Life Span: annual. **Origin**: introduced (from east Asia). **Height**: 1–4 dm. **Stems**: herbaceous, erect or diffuse from a taproot, much branched, sparsely pubescent, retrorsely curved or appressed hairs. **Leaves**: alternate, pinnately 3-foliate, appearing subpalmate; leaflets oblong-obovate, 1–2 cm long, about 1/3 as wide, appressed-ciliate on the margins, otherwise glabrous; petioles 1–3(5) mm long; stipules ovate-lanceolate, striate, 4–6 mm long, brown, scarious, many-nerved, persistent. **Inflorescences**: racemes, 1- to 3-(5-)flowered, in upper axils, the flowers sessile or on pedicels to 2 mm. **Flowers**: chasmogamous and cleistogamous intermixed; calyx tube 1–1.5(2) mm long; lobes 5, about equal, oblong, reticulate, ciliate, about as long as the tube; corolla 5–7 mm long, purple to pink (blue). **Fruit**: legume, obovate, acute, inconspicuously reticulate, 3–4 mm long, 1-seeded; seeds black, often mottled, smooth. n=11.

Other Common Names: striate lespedeza, Japanese lespedeza, Japanese clover

Common lespedeza has escaped from cultivation to roadsides, waste places, fields, upland woods, and rocky open areas. It is best adapted to well-drained, fertile soils. It flowers from July to October. It produces high quality forage for all classes of livestock. It tolerates close grazing and is sometimes cut for hay. Deer and turkeys eat the foliage, while rodents and birds eat the seeds.

Common lespedeza was grown in North America as early as in 1846. Several cultivars have been developed. Scarification improves seed germination. It has limited applications for soil stabilization and landscaping.

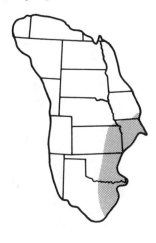

189

Figure 76   *Lespedeza violacea*

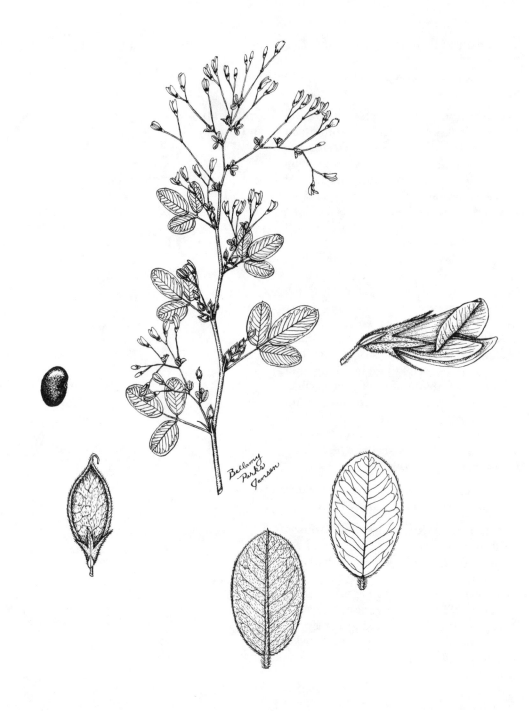

5. *Lespedeza violacea* (L.) Pers.   Violet lespedeza (Figure 76)

[*violaceus* (Lat.): violet, in reference to the color of the flowers.]

Life Span: perennial. *Origin*: native. *Height*: (2)4–8 dm. *Stems*: several from a short-branching caudex, herbaceous, erect or ascending, delicate, much branched above, glabrous to sparsely pubescent, with appressed or ascending hairs. *Leaves*: alternate, pinnately 3-foliate; leaflets elliptic, 1–5 cm long, about 1/2 as wide, glabrous or sparsely pilose above, appressed-pubescent beneath, apex mucronate, base obtuse; petioles slender, nearly as long as the leaflets; stipules filiform, (2)4–6 mm long, persistent. *Inflorescences*: chasmogamous flowers in 4- to 6-flowered racemes, loose, much exceeding the subtending leaves; peduncles long, filiform; cleistogamous flowers in axillary clusters. *Flowers*: calyx lobes 1.5–3 mm long, 2 uppermost connate beyond the middle; corolla purple to violet, 7–10 mm long; banner and wings shorter than the keel; pedicels 2–6 mm long. *Fruit*: legumes, 5–7 mm long, the cleistogamous legumes in clusters, 3–6 mm long, glabrate, 1-seeded; seeds ellipsoid, slightly flattened, 3 mm long, olivaceous to purple or brown, smooth. n=10.

Synonyms: *Lespedeza frutescens* (L.) Britt., *L. prairea* (Mack. & Bush) Britt.
Other Common Name: prairie lespedeza

Violet lespedeza is scattered to locally common in dry or rocky soils of upland woods, prairies, waste ground, and roadsides. It flowers from July to September. It produces good quality forage for livestock and deer. Upland birds eat the seeds.

Commercial seed is seldom available. Germination is improved with scarification. It has the potential to help control erosion, but its landscaping applications are limited.

Two additional low-growing species are similar in appearance and grow in the southeastern Great Plains. *Lespedeza procumbens* Michx., trailing lespedeza, stems are covered with spreading pubescence. *Lespedeza repens* (L.) Bart., creeping lespedeza, plants are trailing and have stipules mostly less than 4 mm in length.

Figure 77 *Lespedeza virginica*

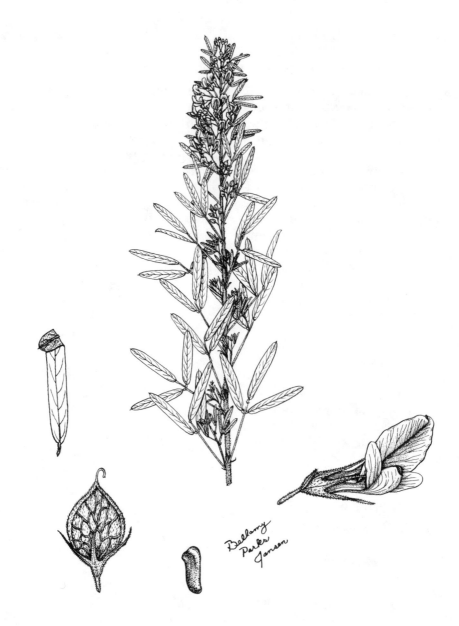

6. *Lespedeza virginica* (L.) Britt.

Slender lespedeza (Figure 77)

[*virginica*: of or from Virginia.]

Life Span: perennial. *Origin*: native. *Height*: 3–10 dm. *Stems*: several from a stout caudex, herbaceous, erect or nearly so, simple or branched above, appressed-pubescent. *Leaves*: alternate, pinnately 3-foliate, crowded, erect to ascending; leaflets linear to narrowly oblong, 6–40 mm long, (3)4–6 times as long as wide, apex obtuse and mucronate, short-strigose (rarely glabrous) above, appressed silky-hairy beneath; stipules filiform, 3–6 mm long, persistent. *Inflorescences*: chasmogamous flowers in few-flowered racemes in upper axils; cleistogamous flowers usually in small axillary clusters; peduncles shorter than the subtending leaves. *Flowers*: calyx lobes of chasmogamous flowers 1.7–3 mm long, the uppermost connate for 1/2 to 2/3 of their length; calyx lobes of cleistogamous flowers shorter; chasmogamous corolla purple to pink, 6–8 mm long; keel occasionally longer than the banner and wings. *Fruit*: legume, thinly and inconspicuously strigose, 4–7 mm long; cleistogamous legume smaller, 1-seeded; seeds green to tan, shiny, ellipsoid, slightly flattened, 2.5–3 mm long. n=10.

Other Common Name: slenderbush lespedeza

Slender lespedeza is scattered to common on prairie, upland woods, river banks, and roadsides. It flowers from May to October. It is grazed by cattle and decreases with continued heavy grazing. Foliage is also eaten by deer and turkeys. Seeds are consumed by upland birds. It is a good honey plant.

Seed is seldom commercially available. Scarification improves germination. It helps to stabilize the soil and is often a pioneer plant on road cuts. It has little potential for landscaping.

Lespedeza stuevei Nutt., tallbush lespedeza, is similar in appearance. Its leaflets are oblong to elliptic or ovate, and they are less than 3 times as long as wide. It also grows in the southeastern Great Plains.

14. LOTUS L.

[*lotos* (Gk.): ancient name applied to several kinds of plants.]

Annual or perennial herbs or suffrutescent plants; leaves odd-pinnately compound with (1)3–5 leaflets; stipules gland-like or obsolete; flowers in umbels or solitary on leafy-bracteate peduncles from the axils; flowers small, yellow or reddish (sometimes white); calyx campanulate or obconic, 5-lobed, lobes nearly equal; corolla papilionaceous, petals clawed, free from stamens; banner ovate or obtuse, not articulate; keel usually fused at both margins, incurved, beaked; stamens 10, diadelphous, filaments partly dilated at the tip; ovary sessile; legumes linear, terete, sessile in calyx, several-seeded, dehiscent.

About 140 species have been described, and nearly 2/3 are native to the Mediterranean region of Eurasia and Africa. About 40 species are native to western North America. Their natural range has been expanded by man. Two species are commonly found in the Great Plains, and a third is occasionally collected.

A. Leaves 3-foliate; stipules reduced to glands; flowers usually solitary, occasionally in pairs; annual . . . . .1. *L. purshianus*
A. Leaves 5-foliate; stipules wanting; flowers in umbels of 3 or more; perennial
. . . . .2. *L. corniculatus*

Figure 78   *Lotus corniculatus*

1. *Lotus corniculatus* L.   Birdsfood trefoil (Figure 78)

[*cornu* (Lat.): horn, referring to the shape of the legume.]

Life Span: perennial. *Origin*: introduced (from Eurasia). *Height*: decumbent branches may be up to 6 dm long. *Stems*: several from a stout crown, prostrate to ascending or rarely erect, taprooted. *Leaves*: alternate, pinnately 5-foliate, sessile or nearly so, lower pair of leaflets basal on the rachis, upper 3 leaflets grouped apically, elliptic to oblanceolate, 5–15 mm long, leaflets about 1/2 as broad as long or broader; stipules wanting. *Inflorescences*: umbels of usually 3–8 flowers, axillary on peduncles that exceed the subtending leaves; pedicels 1–3 mm long (sometimes obsolete). *Flowers*: calyx lobes 5, linear to triangular, about the length of the tube; corolla papilionaceous, yellow to orangish-red, 1.5 cm long; banner as broad as long, exceeding the wings and keel; stamens 10, diadelphous; filaments unequal, the 5 larger dilated at the tip. *Fruit*: legume, 2–4 cm long, terete, straight, brown to black at maturity, 10- to 15-seeded, valves splitting and twisting when mature; seeds 1.5–2 mm long, broadly reniform, olive to black, often mottled. 2n=24.

Other Common Name: birdsfoot deervetch

Birdsfoot trefoil is planted for hay and pasture. It has escaped into roadsides, wasteland, and pastures. It flowers from June to early September. It has had limited use for summer pastures. It does not cause bloat in cattle, probably because it contains condensed tannins that precipitate the soluble leaf proteins that cause bloat. It occasionally produces potentially toxic amounts of a cyanogenic glucoside which can be hydrolyzed by enzymes to produce hydrocyanic acid. There are no reports of actual livestock losses. Upland birds and rodents have been reported to eat the seed.

Over 25 cultivars are commercially available. Seed should be scarified and inoculated with rhizobial bacteria before planting. It has also been used to stabilize gullies, roadsides, and dunes.

Lotus tenuis Waldst. & Kit. *ex* Willd., narrowleaved trefoil, may be occasionally found in the Great Plains. It is similar in appearance to *L. corniculatus*, except that *L. tenuis* leaflets are less than 1/2 as broad as long and are linear to lanceolate.

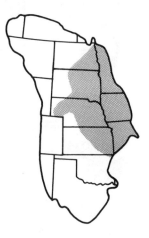

197

Figure 79   *Lotus purshianus*

2. *Lotus purshianus* Clem. & Clem.   American deervetch (Figure 79)

[*purshianus*: named for Frederick Pursh (1774–1820), botanist and author of Flora Americae Septentrionalis (1814).]

Life Span: annual. *Origin*: native. *Height*: 2–8(12) dm. *Stems*: herbaceous, erect, branched, silky-villous to nearly glabrous (especially with maturity). *Leaves*: alternate, nearly sessile, 3-foliate, terminal leaflet short-stalked; leaflets ovate to lanceolate, apex acute, 1–2 cm long, 2–9 mm wide, densely covered with pubescence; stipules gland-like, minute. *Inflorescences*: flowers solitary or rarely paired from the axils, on a short peduncle (about equaling the subtending leaves), usually with a single leaflike bract. *Flower*: calyx 6–8 mm long, hirsute; lobes 5, linear, longer than tube, equaling or exceeding the corolla; corolla white with pink veins (rarely yellowish-white), banner streaked with red; wings oblong, auricled; keel crescent-shaped; stamens 10, about equal, filaments dilated at the tip. *Fruit*: legume, deflexed, 2–3.5 cm long, terete or nearly so, glabrous, dehiscent, usually 4- to 9-seeded; seeds mottled, short oblong, plump, 2.5–3 mm long, 2 mm wide. n=7.

Synonyms: *Lotus americanus* (Nutt.) Bisch., *Acmispon americanus* (Nutt.) Rydb., *Hosackia americanum* (Nutt.) Piper
Other Common Names: prairie trefoil, Spanish clover, Dakota vetch

American deervetch is locally common on sandy soils of dry prairies. It is frequently found on acid soils. It is also prominent along railroads, on abandoned cultivated land, and on overgrazed rangeland. It flowers from June to early September. Domestic livestock eat the foliage, especially early in the season. It has been cultivated, but yields were generally unsatisfactory. Seeds are eaten by upland birds and rodents, and foliage is locally important to deer and pronghorn.

A hot water treatment improves germination of American deervetch, but it is still not high. It has been infrequently used to control erosion in gullies. It has little landscaping value, and seed is seldom commercially available.

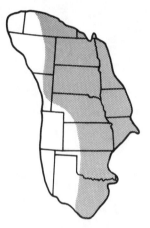

15. LUPINUS L.

[*lupus* (Lat.): wolf, its use for this genus is uncertain, perhaps from an old belief that these plants destroyed the soil.]

Annual or perennial herbs (rarely shrubs); stems erect or ascending, leafy or sub-scapose, from taproots; leaves alternate, palmately compound, with (1)3–15(many) leaflets, petiolate; stipules distinct or partially adnate to the petiole; inflorescences of terminal racemes or spikes; flowers perfect; calyx tube short and asymmetrical, bilabiate, upper lobe slightly bulged to saccate or short-spurred at base; corolla papilionaceous, white, pink, blue, or yellow; sometimes bicolored; banner broad, commonly with reflexed margin; wings commonly connivent by their edges and enclosing the mostly falcate-pointed keel; stamens 10, monadelphous, anthers of 2 kinds, 5 large alternating with 5 smaller; legume somewhat flattened, somewhat constricted between seeds, dehiscent along 2 sutures, 2- to many-seeded.

Primarily a North American genus, but some are native to all continents except Australia. Nearly 200 species have been described. Plants are highly variable making positive identification difficult. Many specimens are intermediate between species descriptions, and hybridization is presumed. Five species are recorded in the Great Plains of which 3 are common.

A. Plants annual; legumes usually 2-seeded ....3. *L. pusillus*
A. Plants perennial; legumes usually with 3 or more seeds
    B. Flowers 1.2–1.4 cm long, bicolored; plants rhizomatous ....2. *L. plattensis*
    B. Flowers 6–12 mm long; plants with a shortly-branching caudex ...1. *L. argenteus*

Figure 80   *Lupinus argenteus*

Bellamy
Parks
Jansen

1. *Lupinus argenteus* Pursh

Silvery lupine (Figure 80)

[*argente* (Lat): silvery, in reference to the shiny leaves.]

Life Span: perennial. *Origin*: native. *Height*: (1)3–6(10) dm. *Stems*: herbaceous, simple to branched, silky pubescent, erect or ascending from a short-branching caudex. *Leaves*: alternate, palmately compound; leaflets 5–10, narrowly lanceolate to oblanceolate, 2–5 cm long, apex acute to obtuse, upper surface dark green, glabrous to strigulose, lower surface silverish-green, pubescent; petioles 2 times or less the length of the blades; stipules adnate to petioles for 1/2 their length. *Inflorescences*: terminal racemes, loose, 5–10(20) cm long. *Flowers*: calyx tube bilabiate, 1.5–2 mm long, sericeous, upper side saccate or bulged at the base, not spurred, upper lobe bidentate, lower lobe entire; corolla papilionaceous, white to pinkish-white, occasionally dark blue, 6–12 mm long, 7–11 mm wide; banner reflexed; keel falcate. *Fruit*: legume, 1–3 cm long, flattened, brown, silky pubescent, 4- to 6-seeded; seeds gray to light brown, ovate, 4–5 mm long, smooth. n=24.

This taxon is highly variable. *Lupinus argenteus* has been divided into several varieties. The two most common are var. *argenteus* which has larger flowers (9–12 mm long) and leaves that usually dry folded, and var. *parviflorus* (Nutt.) C. L. Hitchc. which has smaller flowers (6–9 mm long) and leaves that usually dry flat.

Synonyms: *Lupinus aduncus* Greene, *L. alpestris* A. Nels., *L. decumbens* Torr., *L. floribundus* Greene, *L. laxiflorus* Dougl. *ex* Lindl., *L. parviflorus* Nutt., *L. stenophyllus* Nutt. *ex* Rydb. Over 20 additional synonyms exist.

Other Common Name: perennial lupine

Silvery lupine is scattered to common in prairies, roadsides, and open woods. It flowers from June to August. Palatability is rated as fair for horses, cattle, and sheep. It is classified as a poisonous plant. Legumes and seeds contain quinolizidine and piperidine alkaloids, such as lupanine and sparteine. Greatest losses have occurred when sheep grazed the legumes. Symptoms are labored breathing, followed by coma, and then death from respiratory paralysis. Ingestion of only 0.25% of animal body weight in seeds has caused death. Lupines also cause crooked calf disease. Deformed calves are born to cows that eat lupines when they are 40–70 days pregnant. Anagyrine is the implicated compound. Deer, elk, and pronghorn eat the plant and are seldom poisoned.

Silvery lupine generally has an excellent seed set. Germination is poor unless the seed is scarified. Commercial seed is seldom available, but it does have a potential value for landscaping. Care should be exercised in using it as a landscaping plant, because it has been reported to cause poisoning in children after they consumed a small quantity of seed.

Figure 81   *Lupinus plattensis*

2. *Lupinus plattensis* S. Wats.

Nebraska lupine (Figure 81)

[*plattensis*: named after the Platte River region.]

Life Span: perennial. **Origin**: native. **Height**: 2–5 dm. **Stems**: herbaceous, appressed pubescent, erect or ascending from rhizomes. **Leaves**: alternate, palmately compound; leaflets 5–11, 2–5 cm long, narrowly to broadly oblanceolate or spatulate, glabrous or glabrate above, appressed hairy beneath, ciliate, acute or acuminate; petioles 2–6 cm long. **Inflorescences**: terminal racemes, 6–15(25) cm long. **Flowers**: crowded to separate, calyx more or less gibbous, not projecting backward as a spur or sac, persistent, bifid, lower lobe entire, 7–9 mm long; corolla papilionaceous, 1.2–1.4 cm long, conspicuously bicolored; banner strongly reflexed, blue with a darker spot; wings and keel white or suffused with blue. **Fruit**: legume 2–3.5(5) cm long, densely hairy, 3- to 8-seeded; seeds 6 mm long, 5 mm wide, nearly circular, flat, yellowish-brown to black, smooth. n=24.

Synonyms: *Lupinus glabratus* (S. Wats.) Rydb., *L. ornatus* Dougl. *ex* Lindl., *L. perennis* L. subsp. *plattensis* (S. Wats.) Phillips **Other Common Name**: Platte lupine

Nebraska lupine is infrequent to abundant in sandy soils of prairies, stream valleys, and hills. It is less common in open woods. It flowers from May to August. It is palatable to livestock, but it is poisonous. Lupine poisoning is discussed under *Lupinus argenteus*.

Nebraska lupine has a potential use in landscapes. A similar species *L. subcarnosus* Hook. is the Texas bluebonnet, which has been used for roadside plantings and landscaping for many years.

Two other perennial lupines in the Great Plains are *L. caudatus* Kell., tailcup

lupine, and *L. sericeus* Pursh, silky lupine. Both have 3 or more seeds in the legume. The calyx of *L. caudatus* is spurred, and the pubescence on the banner does not extend to the upper 1/3 of the surface. The calyx of *L. sericeus* is not spurred, and the pubescence on the banner covers at least 2/3 of the surface.

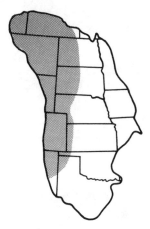

Figure 82 *Lupinus pusillus*

3. *Lupinus pusillus* Pursh Rusty lupine (Figure 82)

[*pusillus* (Lat.): very small or insignificant, in reference to its low habit of growth.]

Life Span: winter annual. *Origin*: native. *Height*: 5–20 cm. *Stems*: herbaceous, loosely villous, erect to decumbent, branching near the base from a winter rosette. *Leaves*: alternate, palmately compound; leaflets 5–9 (occasionally 3 on lowest leaves), 1.5–3.5 cm long, 3–7 mm wide, elliptic-oblong to oblanceolate, often folding along the midvein, glabrous or nearly so above, sparsely pilose below, sometimes ciliate, apex acute to obtuse; petioles 2–5 cm long, broadened and somewhat membranous at the base; stipules 6–7 mm long, adnate to the petioles. *Inflorescences*: racemes, terminating branches, 3–5(7) cm long, usually equaling or exceeding the leaves; peduncles 1–3 cm long, not exceeding the leaves. *Flowers*: calyx villous, bilabiate, tube 2 mm long, upper lip 1.5–2 mm long, lower lip 5–6 mm long; corolla papilionaceous, 8–12 mm long, purple to blue (sometimes white), tinged with pink; banner and wings 8–10 mm long, keel sometimes purple-spotted at the tip. *Fruit*: legumes 2–2.5 cm long (excluding the slender, curved, hairless style), villous-pubescent, usually 2-seeded, constricted between the seeds; seeds light green to brown, mottled with darker brown, nearly circular to obliquely ovate, flattened, 4–5 mm long, 1.5 mm thick. n=24.

Synonyms: *Lupinus intermontanus* Heller, *L. odoratus* Heller
Other Common Name: small lupine

Rusty lupine is common in sandy soils of prairies, badlands, and roadsides. It flowers from May to July. It is poisonous to livestock, at least while seeds are present. Lupine poisoning is discussed under the *Lupinus argenteus* description. Elk, deer, and pronghorn occasionally graze the plant.

Rusty lupine has a limited potential for landscaping. It is easily started from scarified seed.

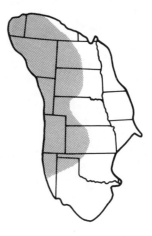

16. MEDICAGO L.

[*medica* (Lat.): for a kind of clover (alfalfa) introduced from Media into Greece.]

Annual or perennial herbs (rarely shrubs) with erect or prostrate stems, stems angled or square, glabrous or pubescent; leaves 3-foliate; leaflets with serrulate margins, terminal leaflet stalked; inflorescences comprised of axillary headlike racemes of small yellow to blue or violet flowers; calyx tube campanulate, 5-lobed; lobes similar but unequal; corolla papilionaceous, banner obovate to oblong, longer than the wings; keel blunt, shorter than the wings; stamens 10, diadelphous; anthers all alike; legume straight or coiled, not enclosed in the calyx, glabrous to spiny, usually indehiscent, 1- to several-seeded.

About 50 species have been described. They are native primarily to the Mediterranean region and extend to Europe and Asia. None are native to North America. Two species are common in the Great Plains, and 2 are relatively rare.

A. Legumes reniform, 1-seeded; peduncles slender, greatly exceeding the subtending leaves; flowers yellow; plants usually annual . . . . .1. *M. lupulina*
A. Legumes coiled, several-seeded; peduncle about equaling the subtending leaves; flowers usually blue to violet; plants perennial . . . . .2. *M. sativa*

Figure 83   *Medicago lupulina*

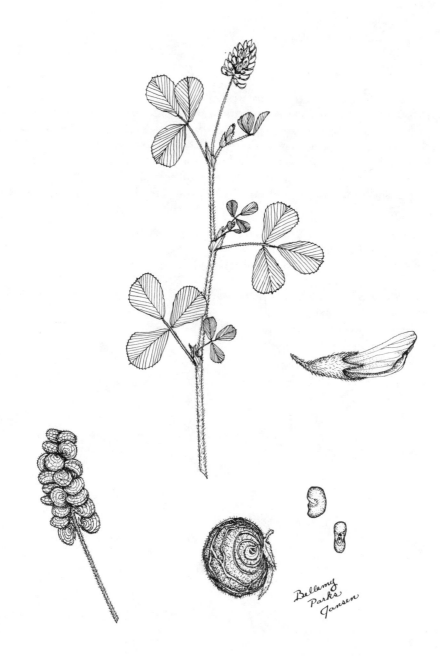

1. *Medicago lupulina* L.
Black medic (Figure 83)

[*lupulinus* (Lat.): little hop, in reference to the hop-like clusters of legumes.]

Life Span: annual (or short-lived perennials). *Origin*: introduced (from Europe and western Asia). *Height*: prostrate with stems 1–8 dm long. *Stems*: herbaceous, widely spreading or ascending from taproots, glabrate to villous. *Leaves*: alternate, pinnately 3-foliate; leaflets elliptic to obovate, 1–2 cm long, 5–10 mm wide, minutely toothed, usually apiculate, glabrous to sparsely pubescent; petioles 0–3 cm long; stipules adnate with petiole for 1/4–1/2 of their length; stipules of upper leaves lanceolate, long acuminate, entire, those of lower leaves wing-like, deeply toothed. *Inflorescences*: headlike racemes, globose to short-cylindric, 4–15 mm long, with 10–50 flowers; peduncles slender, exceeding the subtending leaves by 1–4 times. *Flowers*: calyx glabrous to short-pubescent, less than 1/2 the length of the petals, tube about 0.5 mm long, lobes acuminate; corolla papilionaceous, yellow, 2–4 mm long. *Fruit*: legume, nearly black at maturity, 2–3 mm long, reniform, with conspicuous longitudinal veins, 1-seeded; seed olive to brown or black, reniform, 1.5–2 mm long. 2n=16.

Other Common Names: nonesuch, hop medic, yellow trefoil

Black medic is found throughout the Great Plains in meadows, roadsides, lawns, and prairie ravines. It is often a contaminant in clover seed and has escaped to waste places. It flowers from April to November. It is palatable to all classes of livestock and has some value as a pasture plant and winter cover. Bloat is generally not a problem because animals cannot consume enough of the forage. Foliage is eaten by deer and pronghorn, and the seeds are consumed by upland birds and waterfowl.

Black medic readily reseeds. Seed is not commercially available. Its value for erosion control is low because it is generally an annual. It has no value for landscaping. This species is recognized in Ireland as the "true shamrock."

Medicago minima (L.) Bartal., prickly medic or small burclover, resembles *M. lupulina*. It also is an annual with yellow flowers, but it has coiled legumes with many uncinate prickles. It is much less common and is found most frequently in the southeastern part of the Great Plains.

Figure 84   *Medicago sativa*

2. *Medicago sativa* L.

Alfalfa (Figure 84)

[*sativus* (Lat.): that which is sown, referring to the cultivation of this species.]

Life Span: perennial. *Origin*: introduced (from Europe and western Asia). *Height*: 2–10 dm. *Stems*: herbaceous, erect or decumbent, 5–25 from a common base with a deep taproot. *Leaves*: alternate, pinnately 3-foliate; leaflets oblanceolate to narrowly obovate, 1.5–3 cm long, toothed at summit; petioles of primary leaves 1–5 cm long; stipules ovate-lanceolate, toothed, 5–20 mm long, partially adnate with the petioles. *Inflorescences*: subglobose to short-cylindric racemes, 1–3 cm long, 4- to 45-flowered; peduncles erect, about equaling the subtending leaves, 1–3 cm long. *Flowers*: blue to violet (rarely yellowish-green or brownish-yellow); calyx tube 2–3 mm long, 5-lobed; lobes lanceolate, 2–4 mm long; pedicels 2–3 mm long. *Fruit*: legume, coiled in loose spiral of 1–3 complete turns, finely pubescent, glabrous or sparsely hairy, 2- to 12-seeded; seeds reniform, 2–3 mm long, yellowish-brown, smooth. 2n=32.

Other Common Name: lucerne (Europe and Australia)

Alfalfa is cultivated throughout the Great Plains. It sometimes escapes to roadsides and old fields, but it seldom spreads. It grows best on well-drained soils with a pH of 6.5–7. It is chiefly used for hay for all classes of domestic animals. Alfalfa is occasionally pastured or dried and processed into pellets. It is eaten by nearly all herbivores. It commonly causes bloat when pastured. Seeds are eaten by rodents, rabbits, upland birds, waterfowl, and song birds. It is the primary honey plant in North America.

Seed of many cultivars is readily available. Seed should be scarified and inoculated with rhizobial bacteria before planting. It is known to have been cultivated in Persia by 500 B.C. Alfalfa was introduced into eastern North America in 1736.

Alfalfa can be an important species for erosion control. It has no horticultural value.

Medicago falcata (L.) Arcang., yellow alfalfa, is similar in appearance. Its flowers are yellow and 5–9 mm long, and its legume is straight. It is sometimes classified as a subspecies of *M. sativa*, and it hybridizes freely with *M. sativa*.

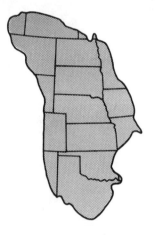

17. MELILOTUS P. Mill.

[*meli* (Gk.): honey; + *lotus* (Gk.): unknown leguminous plant, in reference to it being a source of honey.]

Annual or biennial herbs; leaves alternate, 3-foliate, terminal leaflet stalked; stipules partially adnate to the petiole; inflorescences of elongate racemes, peduncled from upper axils; calyx eventually deciduous, tube campanulate; lobes nearly equal, subulate to lanceolate, acute to acuminate; corolla papilionaceous, small, white or yellow, petals separate; banner oblong to obovate, usually longer than the wings and keel; stamens diadelphous; ovary short, sessile or somewhat stipitate; legume ovate to rotund, glabrous, slightly compressed to nearly globose, 1- to 4-seeded, usually indehiscent.

About 20 species have been described in Europe, Asia, and Africa. None are native to North America. Two species are common in the Great Plains.

Figure 85   *Melilotus officinalis*

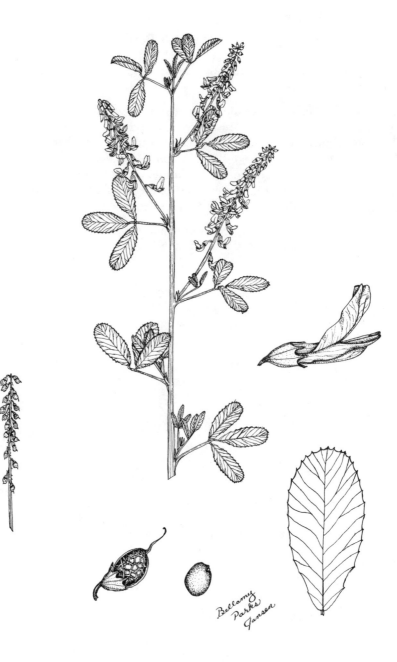

Melilotus officinalis (L.) Pall.   Yellow sweetclover (Figure 85)

[*officina* (Lat.): manufacturing laboratory, indicating the plant was kept for medicinal use by pharmacists.]

Life Span: biennial (rarely annual). *Origin*: introduced (from Eurasia). *Height*: 5–15 dm. *Stems*: herbaceous, erect or ascending, glabrous to sparsely pubescent, much-branched from a taproot. *Leaves*: alternate, pinnately 3-foliate; leaflets oblanceolate to obovate, 1–2.5 cm long, 5–20 mm wide, serrate apically and along the sides, apex obtuse; stalk of center leaflet 3–7 mm long; petioles slender, 1–4 cm long; stipules 6–10 mm long. *Inflorescences*: spikelike racemes, 5–15 cm long (including peduncle), 30- to 70-flowered. *Flowers*: calyx tube 1–2 mm long; lobes deltoid to subulate, 0.5–1 mm long; corolla papilionaceous, yellow (fading with age), 4.5–7 mm long; stamens 10, diadelphous; ovary stalked; pedicels recurved or decurved, 1.5–2 mm long. *Fruit*: legume, ovoid, 2.5–5 mm long, 2–2.5 mm wide, 1.5 mm thick, brown to tan at maturity, cross-veined, short stalked, 1-seeded (rarely 2-seeded); seeds ellipsoid to round, smooth, 2 mm long, yellowish-green to brown. n=8.

Other Common Name: official melilot

Yellow sweetclover is widely naturalized. It was first reported in North America in 1739. It is a weed along roadsides and in waste places and prairies. It flowers from May to October. It is sometimes grown in pastures. It is eaten by most herbivores, and seeds are locally important to wildlife. Domestic livestock must become accustomed to its bitter taste. Sweetclover poisoning occurs in livestock after ingesting moldy hay. Coumarin in sweetclover is converted to dicoumarol during heating and spoilage. Dicoumarol prevents coagulation of blood, and animals may die of internal bleeding. A similar substance is used in rodenticides.

Insects are necessary for pollination. Seed of over 20 cultivars is available. Some of the new cultivars are low in coumarin. Seed should be scarified and inoculated with rhizobial bacteria before planting. It may be used in crop rotations for soil improvement. It is an excellent honey plant.

Yellow sweetclover was recommended by Hippocrates (4th century B.C.) for external treatment of inflamed and swollen body parts and internal treatment of intestinal and stomach ulcers. It has been used as an anticoagulant in more recent times.

Melilotus alba Medic., white sweetclover, is nearly identical except that it has white flowers. It may also be taller (up to 2.5 m) and the flowers are 4–5 mm long. White sweetclover is found throughout the Great Plains.

18. ONOBRYCHIS P. Mill.

[*onobruchis* (Gk.): ancient name of an unknown plant.]

Herbs, sometimes shrubby and spiny; leaves odd-pinnately compound, with many leaflets; stipules scarious; inflorescences of long-peduncled axillary racemes; calyx campanulate, nearly regular, teeth lance-subulate; corolla papilionaceous, pink to purple (rarely white); banner obcordate to obovate; about equal to the keel, much longer than the wings; stamens diadelphous, upper stamens separate at base, but united with others near the middle; legume indehiscent, compressed, ovate to subround, with 1–2 seeds.

About 100 species occur in the Mediterranean region and western Asia. Only 1 has been introduced into the Great Plains.

Figure 86   *Onobrychis vicifolia*

Onobrychis viciifolia Scop.

[*vicia* (Lat.): vetch; + *folium* (Lat.): a leaf, in reference to vetch-like leaves.]

Life Span: perennial. *Origin*: introduced (from western Asia through Europe). *Height*: 3–10 dm. *Stems*: herbaceous, erect, sparingly branched, glabrous to slightly pubescent, ribbed. *Leaves*: alternate, odd-pinnately compound, 10 cm long; leaflets 11–25, linear-oblong to elliptic, 1–3 cm long, with conspicuous venation, usually glabrous on upper surface, sparsely hairy on the margins and midrib beneath, often apiculate or mucronate; stipules small, lanceolate, scarious, midrib conspicuous. *Inflorescences*: spike-like racemes from upper axils, few to several; each flower subtended by a small scarious, lanceolate, bract; peduncles stout, 1–3 dm long. *Flowers*: calyx tube irregular, 2–3 mm long on ·lobed side; lobes lance-subulate, 3–6 mm long, appressed pilose; corolla rose, 8–12 mm long, banner and keel about equal, wings much shorter. *Fruit*: legume, broadly oval, 5–8 mm long, flat, strongly nerved, often with short, blunt spines on the dorsal suture, 1-seeded; seeds broadly reniform, olive to black, 4–7 mm long.

Synonyms: *Onobrychis sativa* Lam. (note: *viciifolia* is sometimes spelled *viciaefolia*)
Other Common Names: esparcet, holy clover

Sainfoin is sparingly planted in the Great Plains. It has occasionally escaped to field margins and road ditches. It is adapted to dry, calcareous soils. It has been used for centuries as a forage crop in Europe and western Asia. Sanfoin does not cause bloat because it contains condensed tannins that precipitate leaf proteins that could otherwise cause bloat. Seeds are eaten by upland birds and domestic fowl. It is the source for the finest European honey.

Several cultivars are commercially available. Nodulation with rhizobial bacteria is not always effective and nitrogen fertilizer may be required for maximum production. Sainfoin has not been successful for erosion control, and it has few landscaping applications.

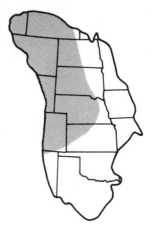

19. OXYTROPIS DC.

[*oxys* (Gk.): sharp; + *tropis* (Gk.): keel; in reference to the characteristic sharp beak of the keel.]

Perennial herbs, acaulescent or nearly so, commonly cespitose, pubescent throughout with the hairs simple and basifixed or dolabriform; leaves and scapes crowded on the crown of a multicipital taproot; leaves odd-pinnately compound, petiolate, often dimorphic; leaflets numerous; stipules strongly or weakly adnate to the petioles; inflorescences of racemes or spikes on long peduncles; flowers papilionaceous, yellowish-white or bright pink to purple; calyx tubular or campanulate, somewhat oblique at the base, 5-lobed; lobes narrow, nearly equal; banner usually erect; wing petals commonly 2-lobed or emarginate at the apex, wings longer than keel; keel abruptly narrowed to a slender, curved or straight beak-like appendage; stamens 10, diadelphous, anthers all alike; legumes erect or spreading, sessile or nearly so, membranous or woody in texture, subspherical to oblong-lanceolate, sometimes inflated, partially bilocular by the intrusion of the ventral suture, dorsal suture not intruded, several- to many-seeded; seeds reniform.

About 200 species occur in circumboreal distribution. They are most numerous at high altitudes. About 40 species grow in North America. Of the 9 in the Great Plains, 4 are common.

A. Pubescent with dolabriform hairs ....2. *O. lambertii*
A. Pubescent with basifixed hairs
    B. Racemes with fewer than 5 flowers; corolla purple to rose-purple; calyx inflated
      in fruit...3. *O. multiceps*
    B. Racemes with 5 or more flowers; flowers white to yellow (sometimes purple
      tipped); calyx not inflated in fruit
        C. Legumes thin, membranaceous at maturity, not fleshy when immature, not
          rigid; flowers generally less than 1.5 cm long................1. *O. campestris*
        C. Legumes woody to coriaceous, fleshy when immature, rigid; flowers usually
          1.5 cm long or more..4. *O. sericea*

Figure 87 *Oxytropis campestris*

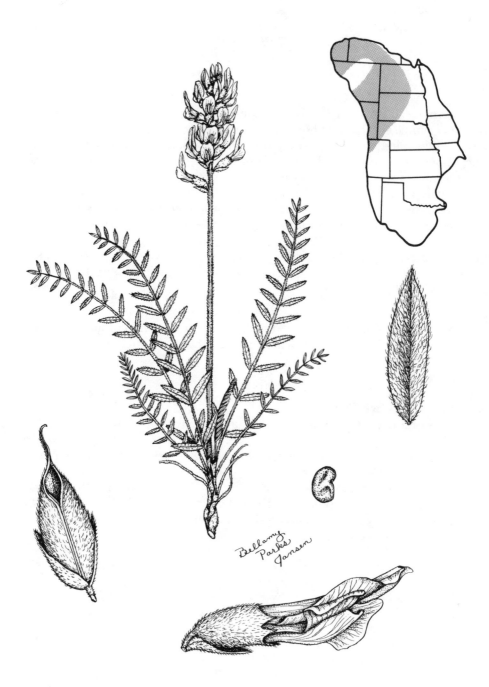

Bellamy
Parks
Jansen

1. *Oxytropis campestris* (L.) DC.   Plains locoweed (Figure 87)

[*campestris* (Lat.): relating to a plain, in reference to the habitat.]

Life Span: perennial. **Origin**: native. **Height**: 2–5 dm. **Stems**: essentially lacking, the leaves and scapes arising directly from a cespitose, much-branched caudex with a taproot. **Leaves**: odd-pinnately compound, dimorphic; primary leaves short with ovate leaflets; secondary leaves with 11–33 leaflets; leaflets oblong to lanceolate, 1–2.5 cm long, with appressed simple hairs above and beneath; petioles pubescent; stipules adnate to the petioles, 5–15 mm long, glabrate or pilose dorsally. **Inflorescences**: capitate or oblong racemes, 8- to 32-flowered. **Flowers**: calyx tube 5–7 mm long, appressed-villous, hairs often black; lobes 5, 1.5–2.5 mm long, appressed-villous, acuminate; corolla papilionaceous, white to cream or yellow (sometimes pink, blue, or purple); banner 1.2–1.5(2) cm long, 6–10 mm wide; wings 1–1.5 cm long, 2–6 mm wide; keel petals may have purple blotches, 1–1.5 cm long, distal appendage to 1 mm long; stamens 10, diadelphous. **Fruit**: legume, nearly sessile, erect to spreading, 1.5–2 cm long including beak to 5 mm long, pubescent, some hairs black, suture strongly intruded so that the legume is nearly bilocular, dehiscing from the tip, thin, membranaceous (rarely coriaceous), many-seeded; seeds 2–2.5 mm long, yellow or brown to black, smooth, reniform. n=16.

The most common variety is *gracilis* (A. Nels.) Barneby. According to McGregor in the *Flora of the Great Plains,* some North Dakota plants have white to purple flowers and are recognized as var. *dispar* (A. Nels.) Barneby.

Synonyms: *Oxytropis dispar* (A. Nels.) K. Schum., *O. gracilis* (A. Nels.) K. Schum., *O.* *macounii* (Greene) Rydb., *O. villosa* (Rydb.) K. Schum.

Other Common Names: slender locoweed, yellow locoweed, late locoweed, plains crazyweed

Plains locoweed is scattered to locally abundant in prairies, meadows, and woods. It is most abundant in gravelly or rocky soils. It flowers from May to July. It is generally unpalatable to livestock.

Commercial seed is not available, although it could be an attractive addition to landscapes. Scarification will improve germination.

Three other species of crazyweed are less common on the Great Plains. *Oxytropis splendens* Dougl. *ex.* Hook., showy crazyweed, has reddish-purple flowers, and some of the leaflets are fascicled on one side of the rachis. *Oxytropis besseyi* (Rydb.) Blank., red crazyweed, has pink to reddish-purple flowers, bracts are rhombic-lanceolate, and the surface of the calyx is not obscured by the pubescence. *Oxytropis lagopus* Nutt. var. *atropurpurea* (Rydb.) Barneby, hare's crazyweed, also has reddish-purple flowers. Its bracts are lanceolate, and the surface of the calyx tube is obscured with pubescence. All three of these species grow in the northwestern Great Plains.

Figure 88 *Oxytropis lambertii*

2. *Oxytropis lambertii* Pursh

Lambert crazyweed (Figure 88)

[*lambertii*: named for Aylmer Bourke Lambert (1761–1842), English botanist.]

Life Span: perennial. *Origin*: native. *Height*: 1–3(5) dm. *Stems*: essentially lacking, the leaves and scapes arising directly from a cespitose caudex with a taproot, 1-several clusters, pubescent with dolabriform hairs. *Leaves*: odd-pinnately compound, erect or ascending, dimorphic, principal leaves 4–20 cm long, with 7–19 leaflets; leaflets linear to narrowly oblong, sometimes nearly orbicular, base and apex acute, 5–40 mm long, 2–7 mm wide, thinly strigose-canescent, with dolabriform hairs; stipules adnate to the petioles. *Inflorescences*: terminal racemes, elevated above the leaves on scapes, 5- to 25-flowered, spreading to erect, 4–12 cm long, elongating in fruit. *Flowers*: calyx tube campanulate, 5–9 mm long, densely villous, 5-lobed; lobes triangular-subulate, 2–4 mm long; corolla papilionaceous, purple to rose or blue (white not uncommon); banner somewhat reflexed, 1.5–2.5 cm long; wings 1.2–2 cm long, enveloping the keel; keel 1–2 cm long, appendage 5–25 mm long; stamens 10, diadelphous. *Fruit*: legume sessile or nearly so, erect or spreading, oblong to cylindric, 5–6 mm wide, 3 cm long including the prominent straight or divergent beak (beak 3–7 mm long), strigose-silky or strigose, soon glabrous, suture intruded about 1/2 across, dehiscing from the top, many-seeded; seeds brown, 2 mm long, smooth broadly reniform to nearly orbicular. n=24.

This is a highly variable species. It hybridizes with others, especially with *O. sericea*.

Synonyms: *Oxytropis hookeriana* Nutt., *O. involuta* (A. Nels.) K. Schum., *O. patens* (Rydb.) A. Nels., *O. plattensis* Nutt., *Aragallus articulata* (Greene) Barneby

Other Common Names: Lambert locoweed, purple locoweed, whitepoint locoweed, rattleweed, sunkta pejuta (Lakota)

Lambert crazyweed is common on prairie uplands. It is most abundant on drier sites. It flowers from April to August. It is generally not palatable to livestock, but they will eat it if other forage is not available.

Lambert crazyweed causes loco disease in cattle, sheep, goats, and especially horses. Loco means "crazy" in Spanish. Large amounts of the plant material must be consumed before poisoning occurs. Animals may have to eat the plants for a few weeks before the indolizidine alkaloids accumulate to the point of being toxic. Cattle and sheep show signs of toxicity after eating 90% of their body weight of Lambert crazyweed. Death occurs with ingestion of 3 times their body weight of the plants. Horses only have to eat 30% of their body weight of the plants to acquire a lethal dose. Animals develop a craving for the plant and graze it in preference to other species. Symptoms of poisoning include crazy actions, running into objects, depression, trembling, and paralysis.

Seeds are not commercially available. Lambert crazyweed has the potential to become a landscape plant.

Figure 89   *Oxytropis multiceps*

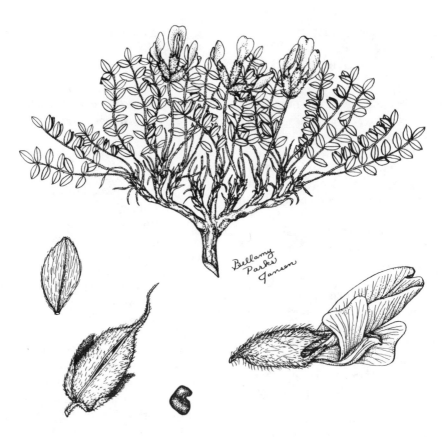

3. *Oxytropis multiceps* T. & G.   Manyhead crazyweed (Figure 89)

[*multus* (Lat.): many; + *ceps* (Lat.): head, in reference to the many inflorescences.]

Life Span: perennial. ***Origin***: native. ***Height***: 2–10 cm. ***Stems***: very short, densely cespitose from a caudex and taproot, the plants forming silvery mounds. ***Leaves***: odd-pinnately compound, 1–5 cm long, with 5–17 leaflets; leaflets 3–12 mm long, lanceolate to oblong, densely silky with appressed hairs; petioles 5–30 mm long; stipules adnate to the petioles. ***Inflorescences***: racemes, terminal on scapes, 1- to 4-flowered, 1.5–3 cm long, ascending or prostrate. ***Flowers***: calyx tube 6–10 mm long, becoming inflated in fruit (8–18 mm long and 5–9 mm in diameter), silky-villous, membranaceous, reddish, 5-lobed; lobes unequal, 2–3 mm long; corolla papilionaceous, purple to rose-purple (drying light blue); banner 1.7–2.5 cm long, 7–9 mm wide, oblong-obovate, emarginate; wings distally widened, 4.5–5.5 mm wide near apex; keel 1.3–1.8 cm long, distal appendage straight or curved and 0.5–1.5 mm long. ***Fruit***: legumes, erect or spreading, ovoid to ellipsoid, wholly or partially included in the inflated calyx, few-seeded, contracted into a beak, papery, not rigid, short-villous, stipe 1.5 mm long; seeds reddish-brown, often mottled with purple, smooth, reniform.

Other Common Name: dwarf locoweed

Manyhead crazyweed is infrequent to locally common on prairie uplands and ridges and on open wooded hillsides. It is unpalatable to livestock, but it can cause loco poisoning (see discussion under *O. lambertii*).

    Commercial seed is not available, although it could be an attractive addition to landscape plantings. Scarification will improve germination.

Oxytropis nana Nutt., dwarf crazyweed, has rigid and coriaceous legumes. The flowers are purple or white with purple spots. It is found only in central and east-central Wyoming.

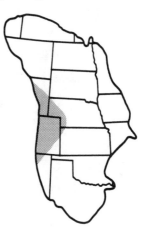

Figure 90   *Oxytropis sericea*

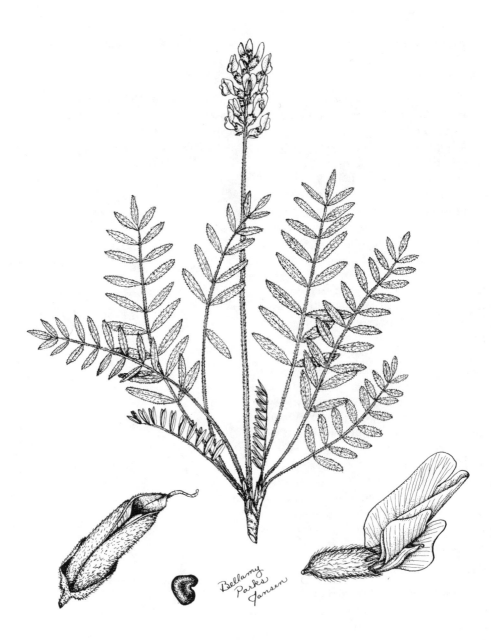

4. *Oxytropis sericea* Nutt. — Whitepoint crazyweed (Figure 90)

[*serikos* (Gk.): silken, in reference to the pubescence.]

Life Span: perennial. **Origin**: native. **Height**: 1–5 dm. **Stems**: essentially lacking, leaves and scapes arising from cespitose much-branched woody caudex with a taproot, scapes erect or ascending, silky-pilose with basifixed hairs. **Leaves**: odd-pinnately compound, 4–30 cm long, erect or ascending, dimorphic, with (7)11–25 leaflets; leaflets elliptic to oblong, 5–35 mm long, 2–9 mm wide, apex acute or obtuse, base acute, gray or silvery with appressed, basifixed hairs; petiole 1–15 cm long; stipules 5–25 mm long. **Inflorescences**: terminal racemes or scapes, usually elevated above the leaves, 5–10 cm long, 6- to 30-flowered. **Flowers**: calyx tube campanulate to tubular, 5-lobed, appressed silky-hairy, with some black hairs present; lobes acuminate, unequal, 2–5 mm long; corolla papilionaceous, white (fading to yellow), with purple-tipped keel; banner 1.5–2.5 cm long, 8–10 mm wide, with a broad claw, blade oblong-ovate, emarginate or deeply lobed; wings 1.5–2 cm long, blade 8–14 mm long, widening distally, 5–9 mm wide near emarginate apex; keel 1.2–1.8 cm long, distal curved or straight appendage 1–2 mm long. **Fruit**: legumes erect, oblong or ovoid-oblong, short-pilose to strigose, woody to coriaceous, 1–2.5 cm long, longer than the calyx, 5–8 mm in diameter, with a short beak, many-seeded; seeds brown, 2–2.5 mm long, smooth, reniform.

Synonyms: *Oxytropis pinetorum* (Heller) K. Schum., *O. spicata* (Hook.) Barneby
Other Common Names: silky crazyweed, white locoweed

Whitepoint crazyweed is infrequent to common on prairie uplands, stream banks and valleys, and open wooded hillsides. It is most common on sandy, gravelly, or rocky soils. It flowers from April to June. It is generally unpalatable to livestock, but it can cause loco disease (see discussion under *O. lambertii*).

Commercial seed is not available. Scarification will improve germination. It has potential as a landscape plant.

Oxytropis deflexa (Pall.) DC. var. *sericea* T. & G., pendulouspod crazyweed, is shortly caulescent, with blue to purplish flowers, and the stipules are adnate to the petioles for only 1–3 mm.

20. PSORALEA L.

[*psoraleos* (Gk.): scurfy, referring to the many glands.]

Perennial herbs, sometimes shrubs, from rootstalks or tuber-like roots; leaves alternate, palmately compound (rarely pinnately compound), 3- to 7-foliate, sometimes single, usually glandular-punctate; leaflets entire; petioles well developed; stipules usually persistent; inflorescences consisting of peduncled spikelike racemes, terminal or from upper axils; calyx campanulate, often gibbous, sometimes oblique, persistent, 5-lobed; lobes equal or lower lobe longer; corolla papilionaceous, blue to purple (rarely white); banner usually clawed; wings usually equal to the banner; keel shorter and incurved; stamens 10 (rarely 9), diadelphous or sometimes monadelphous; anthers uniform; ovary sessile or nearly so; legume short, ovoid, flattened or turgid, usually indehiscent, sometimes terminated by the indurate and persistent style, 1-seeded; seeds ellipsoid to ovoid.

Over 150 species have been described, occurring in North and South America, Africa, and Australia. About 50 species are found in North America, especially in the southern and southwestern portions. Six are common to the Great Plains, and 2 more are occasionally reported.

A. Plants low, stemless or stems 1–4 dm tall; from tuberous or thickened roots; petioles
   longer than the leaflets. . . . . .4. *P. esculenta*
A. Plants taller, stems leafy and much branched; lacking tuberous or thickened roots;
   petioles shorter than or equaling the length of the leaflets
   B. Flowers 1.5–2 cm long, in dense spikelike racemes; leaves mostly 5-foliate, rarely
      7-foliate. . . . . .2. *P. cuspidata*
   B. Flowers less than 1.5 cm long, in racemes; leaves 3- to 5-foliate
      C. Leaves 3-foliate, leaflets sessile; plants glabrous to sparsely strigose; inflorescences scarcely longer than the leaves. . . . . .5. *P. lanceolata*
      C. Leaves 3- to 5-foliate, at least lower leaves 5-foliate; inflorescence obviously
         longer than the leaves
         D. Plants densely silvery-hairy throughout; spike short, interrupted
            . . . . .1. *P. argophylla*
         D. Plants greenish to gray, leaflets glabrous to sparsely strigose on upper surface
            E. Calyx large, inflated and enclosing the legume at maturity; leaves mostly
               5-foliate; flowers in whorls of 3–6(10) at each node. . . . . . . .3. *P. digitata*
            E. Calyx small, much shorter than corolla or legumes; leaves mostly 3-foliate; flowers 1–3(4) at each node . . . . .6. *P. tenuiflora*

233

Figure 91   *Psoralea argophylla*

234

1. *Psoralea argophylla* Pursh

Silverleaf scurfpea (Figure 91)

[*argos* (Gk.): shining; + *phyllon* (Gk.): leaf, in reference to the silvery hairs on the herbage.]

Life Span: perennial. ***Origin***: native. ***Height***: 2–6(8) dm. ***Stems***: herbaceous from adventitious buds on woody roots and from creeping rhizomes, much branched, erect or ascending, sometimes flexuous, densely white-silky. ***Leaves***: alternate, 5-foliate (sometimes 4-foliate) on main stem and palmately 3-foliate on branches; leaflets elliptic or lanceolate to narrowly obovate, 1–5 cm long, 6–18 mm wide, densely white-silky on both sides, but less pubescent above and more green above, apex obtuse or acute, usually with short mucro; petioles shorter than to equaling the leaflets, 1–3 cm long; stipules of lower nodes 1–2 cm long, shorter above. ***Inflorescences***: axillary spikes, 2–8 cm long, 1–5(8) separated whorls, each whorl with 2–8 flowers; bracts slightly exceeding the flowers. ***Flowers***: sessile, 8–10 mm long, calyx persistent, tube campanulate, 2–5 mm long, densely white-silky; lowest lobe attenuate, 8–10 mm long, twice as long as the upper 4 lobes; corolla papilionaceous, dark-blue to purple, fading to yellow or brown; banner obovate 5–7 mm long, wings 4–6 mm long, keel 4–5 mm long. ***Fruit***: legumes, ovoid, silky, 6–7(9) mm long excluding a short beak, pericarp rigid, 1-seeded; seeds olive to black, 4–5 mm long, orbicular to reniform, smooth. n=11.

Synonyms: *Psoralidium argophyllum* (Pursh) Rydb., *P. collinum* Rydb.
Other Common Name: ticanicahn (Lakota)

Silverleaf scurfpea is scattered to locally abundant on dry prairies, hills, and woodlands. It is most common in sandy or rocky soils. It flowers from June to September. It is of little value to domestic livestock. Plains pocket gophers eat the roots. Deer and pronghorn occasionally consume the foliage. Seeds may be poisonous. One case of severe poisoning in a child was reported after he ate a relatively large quantity of seeds.

The plants break off near the soil surface in early fall. Seed is scattered as the plant is tumbled by the wind. Seed set is poor. Stratification and scarification improve germination. Seed is not commercially available. It has little potential for landscaping.

Great Plains Indians used silverleaf scurfpea to treat wounds, and a mild stimulant was made from the roots. Lakota made baskets from the green stems. They also fed the roots to horses when they were tired.

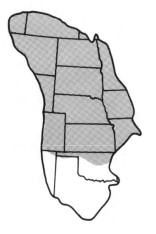

Figure 92   *Psoralea cuspidata*

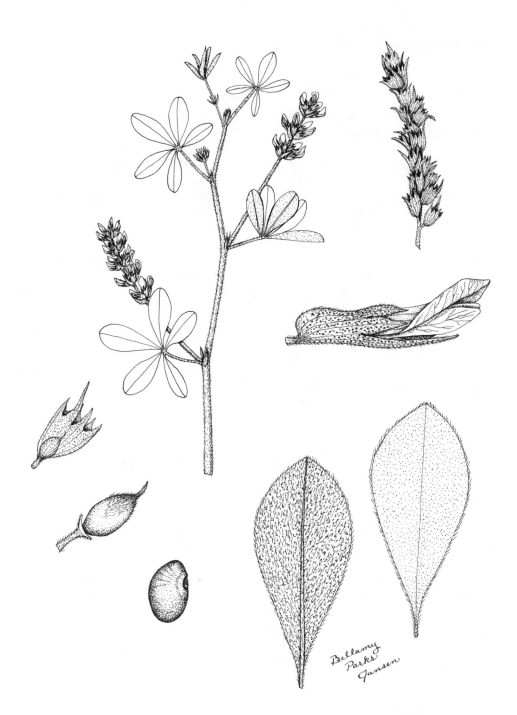

2. *Psoralea cuspidata* Pursh

Largebracted scurfpea (Figure 92)

[*cuspidatus* (Lat.): pointed, in reference to the calyx lobes.]

Life Span: perennial. *Origin*: native. *Height*: 1–6(9) dm. *Stems*: herbaceous from elongated fusiform or ellipsoidal taproot, erect or ascending, much branched, stout, thinly strigose to glabrate. *Leaves*: alternate, palmately 5-foliate (rarely 7-foliate, upper leaves sometimes 3-foliate); leaflets 1.5–4(6) cm long, center leaflet longest, 5–20(30) mm wide, elliptic to narrowly obovate or oblong-obovate, glandular-punctate, glabrous above, strigose beneath, margins pubescent, apex acute to obtuse, sometimes with a mucro; petioles 1–5 cm long; stipules of lower nodes 1–2 cm long, ovate; upper stipules lanceolate, 1 cm long. *Inflorescences*: racemes, dense, thick-cylindric, axillary spikelike 2–6(9) cm long, 3.5 cm wide; peduncles longer than the petioles; bracts 1–1.5 cm long. *Flowers*: calyx persistent, gibbous at base, 8–12 mm long, strigose, densely glandular-punctate; upper 4 lobes 3–6 mm long, lowest lobe 7–12 mm long; corolla papilionaceous, blue or purple (rarely white); banner 1.5–2 cm long, claw 4–6 mm long; wings 1 cm long, claws 5–7 mm long; pedicels short. *Fruit*: legume, ovoid, 6–8 mm long, thin-walled, papery, glandular, 1-seeded; style persistent, 2 mm long; seeds olive to gray or brown, mottled, slightly flattened, round to broadly elliptic, 4–5 mm long, 3–3.5 mm wide, shiny. n=11.

Synonym: *Pediomelum cuspidatum* (Pursh) Rydb.
Other Common Name: tallbread scurfpea

Largebracted scurfpea is infrequent to common on prairie uplands, especially in rocky or sandy soil. It flowers from May to July. Cattle rarely eat the plants, but they still rapidly decrease with grazing. Deer and pronghorn utilize the foliage, while rodents and upland birds eat the seeds.

Indians peeled and ate the root either raw or cooked. The root was dried and ground or pounded into flour.

Seeds are not commercially available. It has little value for landscaping.

237

Figure 93   *Psoralea digitata*

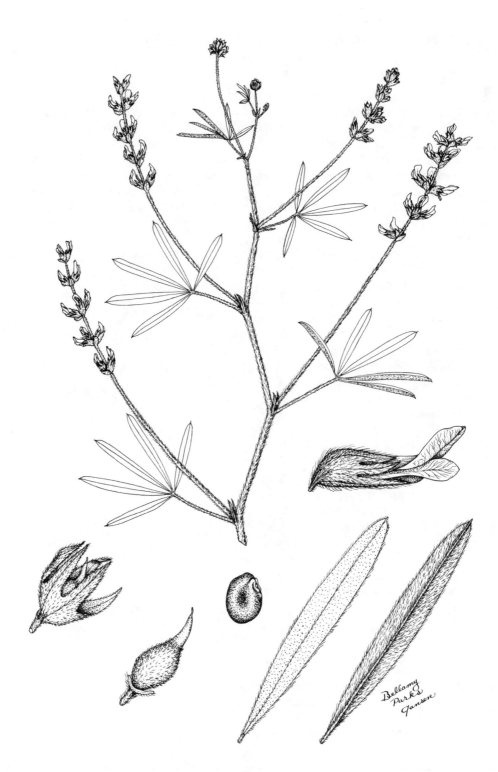

3. *Psoralea digitata* Nutt.

[*digitatus* (Lat.): having fingers, in reference to the palmately compound leaves.]

Life Span: perennial. **Origin**: native. **Height**: 3–8(10) dm. **Stems**: simple below, branched above, from rhizomes or thick (not fleshy) taproots, appressed canescent. **Leaves**: alternate, usually 5-foliate (sometimes 7-foliate) at least on the main stem and 3-foliate on some branches; leaflets linear to oblong-linear, 1.5–6 cm long, middle leaf longest, 3–7 mm wide, obtuse or mucronate, smooth to glandular above and strigose on the midvein, densely strigose beneath; petioles 2–7 cm long; stipules lanceolate, reflexed, involute on drying, strigose, 3–8(10) mm long. **Inflorescences**: axillary, spikelike racemes, few-flowered, interrupted, flowers in whorls of 3–6(10); peduncles much longer than the leaves (5–16 cm long). **Flowers**: nearly sessile; calyx tube campanulate, 2.5–3.5 mm long at anthesis (exceeding fruits when mature), strigose; upper 4 lobes 3–4 mm long, lower lobe slightly longer, acute to acuminate; corolla papilionaceous, 7–10 mm long, blue or purple (rarely white); banner obovate, 7–9 mm long; wings 6–7 mm long. **Fruit**: legumes ovoid, thin-walled, papery, 6–8 mm long, tapering to a flat beak; seeds olive to gray or brown, broadly elliptic, shiny, slightly flattened, 4–4.5 mm long, 3 mm wide. n=11.

Synonym: *Psoralidium digitatum* (Nutt.) Rydb.
Other Common Name: palmleaved scurfpea

Finger scurfpea is infrequent to locally common on prairies and open wooded hillsides. It is most abundant in sandy soils. It flowers from May through July. It is apparently unpalatable to domestic livestock, although deer utilize it. Rodents consume the seeds.

Seeds are not commercially available. Germination is improved with scarification and stratification. Finger scurfpea has little landscaping potential.

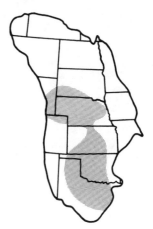

Figure 94   *Psoralea esculenta*

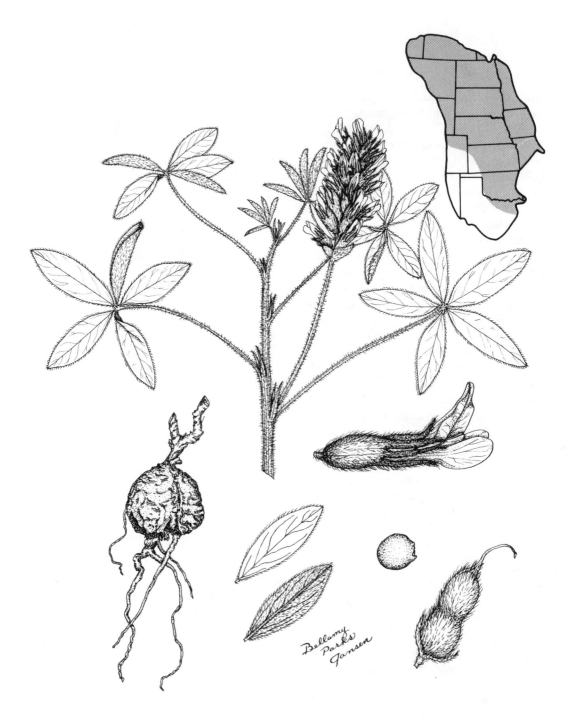

Bellamy
Parks
Jansen

4. *Psoralea esculenta* Pursh
Breadroot scurfpea (Figure 94)

[*esculentus* (Lat.): edible, in reference to the root.]

Life Span: perennial. *Origin*: native. *Height*: 1–4 dm. *Stems*: herbaceous, erect, usually solitary (up to 3) with axillary peduncles giving impression of lateral branches, densely villous-hirsute. *Roots*: taproots, fusiform to subglobose, 5–10 cm long, 1.5–5 cm thick, surface brown and leathery, white and starchy inside. *Leaves*: alternate, sometimes clustered; palmately compound, 5-foliate (sometimes 3-foliate); leaflets 2–6 cm long, 6–16 mm wide, middle leaflet longest, elliptic or oblanceolate to obovate, apex obtuse or acute (sometimes with a mucro), glabrous above, with long appressed hairs and occasionally glandular dots beneath, ciliate; petioles (2)5–10(15) cm long, villous-hirsute; stipules conspicuous, oblong to lanceolate, 1–2 cm long. *Inflorescences*: terminal racemes, few, dense, (2)4–6(8) cm long, 2–2.5 cm wide. *Flowers*: calyx 7–10 mm long, persistent or not, 5-lobed; lobes nearly equal or lowest lobe slightly longer than the others; corolla papilionaceous, blue to purple (rarely tinged with white) at anthesis, fading to yellow, drying to brown, 1.5–2 cm long at flowering (expanding with maturity), hirsute (rarely glabrous). *Fruit*: legume, thin, papery, hirsute, 1- to 2-seeded, 2 cm long, long-beaked; seeds gray to brown, nearly round to broadly elliptic or narrowly ovate, plump to slightly flattened, 4–6 mm long, 3–4.5 mm wide. n=11.

Synonym: *Pediomelum esculentum* (Pursh) Rydb.

Other Common Names: prairie turnip, Indian turnip, Indian breadroot, prairie potato, pomme de prairie, esharusa (Crow), tipsinnah (Lakota), patsuroka (Pawnee), nugthe (Omaha-Ponca), tdokewihi (Winnebago)

Breadroot scurfpea is infrequent to common in dry prairies, bluffs, valleys, and open woodlands. It does not occur in dense stands. It blooms from May through July. In late summer the plant breaks off near the soil surface and scatters seed as it tumbles with the wind. It has little forage value for cattle, but it is fair for sheep and wildlife. It increases with light grazing and rapidly disappears with continued moderate to heavy use.

Small quantities of seed are occasionally commercially available. Seed germination is enhanced with stratification. Breadroot scurfpea has little value as a landscape plant.

Breadroot scurfpea was an important food for the Great Plains Indians, who ate both raw and cooked roots of the plant. Fresh roots were boiled or roasted. Roots were sometimes dried and stored for long periods. Dried roots were pulverized, mixed with water, and baked over coals.

Psoralea hypogaea Nutt., little breadroot, has a similar appearance. It is stemless or has very short stems (1–2 cm long). Leaves are 5- to 7-foliate and linear-lanceolate to linear oblong. Its corolla is less than 1.5 cm long. It is found primarily in the southwestern portion of the Great Plains.

Figure 95   *Psoralea lanceolata*

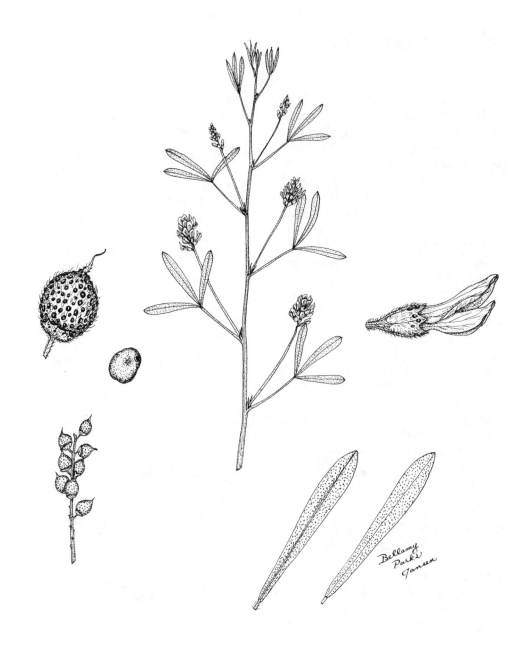

5. *Psoralea lanceolata* Pursh

Lemon scurfpea (Figure 95)

[*lanceolatus* (Lat.): spear-like, in reference to the shape of the leaflets.]

Life Span: perennial. *Origin*: native. *Height*: 1–6(8) dm. *Stems*: herbaceous, erect or ascending, much branched, glabrous to sparingly strigose, glandular-punctate, from long (up to 10 m) rhizomes. *Leaves*: alternate, palmately 3-foliate (occasionally 5-foliate); leaflets 1.5–5 cm long, 2–13 mm wide, variable, linear to narrowly oblanceolate, acute to mucronate, glandular-punctate, glabrous to sparsely strigose below; petioles mostly equaling or shorter than the leaflets, 1–2 cm long; stipules linear-lanceolate, 3–10 mm long. *Inflorescences*: axillary racemes, 1–3 cm long; peduncles 2–5 cm long, scarcely projecting above the foliage. *Flowers*: calyx tube 2–3 mm long, campanulate, glandular, sparsely strigose, lobes nearly equal, 2 united and shortly bifid, broadly triangular, less than 1/2 as long as the tube; pedicels 0.5–3 mm long; corolla papilionaceous, white or violet-tinged, 5–7 mm long; banner orbicular, 5–6 mm long, claw 1 mm long; wings oblong-oblanceolate, 3–4 mm long, claw 1 mm long. *Fruit*: legume, globose, 4–6 mm long, abruptly short-beaked, glabrate to densely-villous, glandular, 1-seeded; seed reddish-brown, nearly round in outline, slightly flattened, 3–5 mm long, 3–4.5 mm wide. n=11.

Synonyms: *Psoralidium lanceolatum* (Pursh) Rydb., *P. micranthum* (A. Gray) Rydb., *P. stenophyllum* (Rydb.) Rydb.

Lemon scurfpea is common on dry prairies and hills, especially in sandy soils. It is especially common in sand hill blowouts. It flowers from May through August. It is not grazed by domestic livestock or wildlife. Rodents utilize the seeds.

Lemon scurfpea can be important for erosion control. It is not planted, but its rhizomatous growth and low palatability make it excellent for binding sandy soil. Seed is not commercially available, and it has no landscaping value. Bruised foliage has a lemon fragrance.

Psoralea linearifolia T. & G., narrowleaf scurfpea, is similar in appearance. It is not as common as lemon scurfpea and is found primarily in the southwestern portions of the Great Plains. Narrowleaf scurfpea has flowering racemes that are much longer (2–8 cm long) than the leaves. The lower leaves are 3-foliate, and the upper leaves are single or 2-foliate. Plants are 3–7 dm tall, and the leaflets are linear.

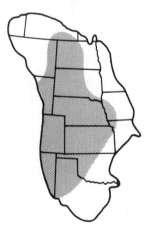

243

Figure 96   *Psoralea tenuiflora*

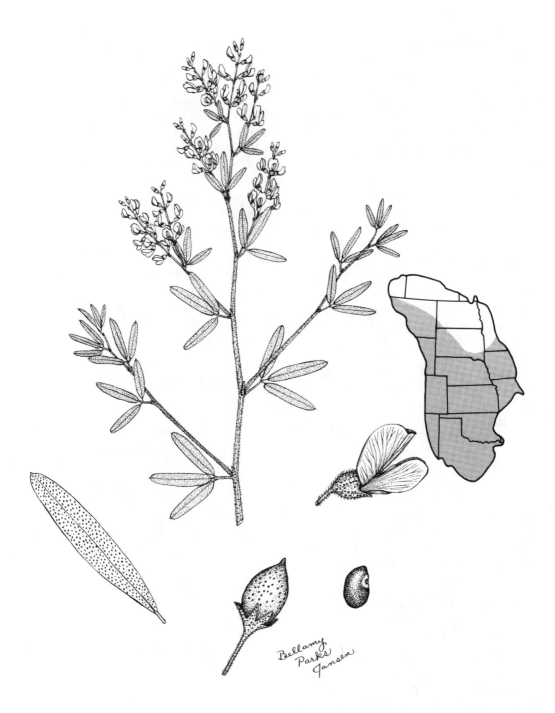

6. *Psoralea tenuiflora* Pursh

Slimflowered scurfpea (Figure 96)

[*tenuis* (Lat.): thin, slender; + *floris* (Lat.): flower, referring to the slender flowers.]

Life Span: perennial. *Origin*: native. *Height*: 3–10(13) dm. *Stems*: herbaceous, erect from taproot (rarely rhizomatous), much branched, hoary pubescent when young, strigose later, glandular-dotted. *Leaves*: alternate, palmately compound, 3-foliate (5-foliate on lower leaves which tend to be caducous); leaflets linear-oblanceolate to obovate, 1–5 cm long, 4–12 mm wide, obtuse and mucronate at the apex, glabrate above, gray-strigose beneath, glandular-dotted on both sides, less so beneath; petioles mostly shorter than the leaflets, 4–20 mm long; lower stipules deltoid, 6–9 mm long; upper stipules lanceolate, 2–3 mm long. *Inflorescences*: axillary racemes, 2–6(10) cm long, loose, with 1–3(4) flowers at each node; peduncles longer than the leaves; bracts ovate to lanceolate, 2–6 mm long. *Flowers*: calyx tube campanulate, 1.5–2.5 mm long or more, glandular-punctate, strigose or loosely villous-hirsute; lobes acuminate, lower lobe 1.2–2.5 mm long, upper lobes 1–1.5 mm long; corolla papilionaceous, blue to purple (rarely white); banner obtuse-obovate, 4–6(8) mm long; wings 3–6 mm long; pedicels 1.5–4 mm long. *Fruit*: legumes 5–9 mm long, flattened, obovate or oblong, short beaked, glabrous, densely glandular-dotted, often asymmetric; seeds grayish-green to orangish-brown, sometimes purple spotted, reniform to nearly round, slightly flattened, shiny, 4–5.5 mm long. n=11.

Two varieties are recognized. Var. *tenuiflora*, slimflowered scurfpea, grows in the central and western Great Plains. Var. *floribunda* (Nutt.) Rydb., manyflowered scurfpea, grows in the east central and southeastern Great Plains. Var. *floribunda* has larger flowers (6–8 mm long), 2–4 flowers at each node, and longer racemes (6–10 cm) with crowded flowers. Its calyx tube is over 2.5 mm long and loosely villous-hirsute, rather than strigose. Where their ranges overlap, all gradations occur.

Synonyms: *Psoralidium tenuiflorum* (Pursh) Rydb., *P. batesii* Rydb., *P. floribundum* (Nutt.) Rydb.
Other Common Names: wild alfalfa, few-flowered scurfpea, scurfy pea, slender scurfpea, ticanicahu tanka (Lakota)

Slimflowered scurfpea is common on dry prairies, rocky banks and openings in woodland. It flowers from May through September. Palatability is low, and it is generally not grazed by domestic livestock. After curing in hay, it is readily eaten. It has been reported to be toxic to horses and cattle, but there is no substantiated evidence. It is an excellent honey plant, and its seeds are eaten by wildlife. The stems break off near the soil surface in late summer to early fall. The wind tumbles the plant along the ground scattering seed.

Seed is not commercially available. Slimflowered scurfpea has little value for landscaping or erosion control.

Lakota made tea from the roots for headache, burned the plant to repel mosquitoes, and made garlands from the tops to be worn for protection from the sun. Other tribes reportedly used slimflower scurfpea for fish poison.

21. ROBINIA L.

[*Robinia*: named after Jean Robin (1550–1629) and his son Vespasian Robin (1579–1662), herbalists and botanists for the King of France.]

Trees or shrubs with odd-pinnately compound leaves; stipules setaceous or modified into spines; inflorescences of axillary racemes with large flowers; calyx tube hemispheric or broadly campanulate, 5-lobed; lobes bilabiate, lower 3 about equal, upper 2 connate for 1/3 or more of their length; corolla white, pink or purple; banner suborbicular, more or less reflexed; wings obliquely obovate, long-clawed; keel strongly upwardly curved, long-clawed; stamens 10, diadelphous; ovary elongate; legumes elongate, flat, many-seeded.

About 20 species occur in North America. Two are found in the Great Plains. One has been planted and rarely escapes. The other one is common and is described here.

Figure 97 *Robinia pseudoacacia*

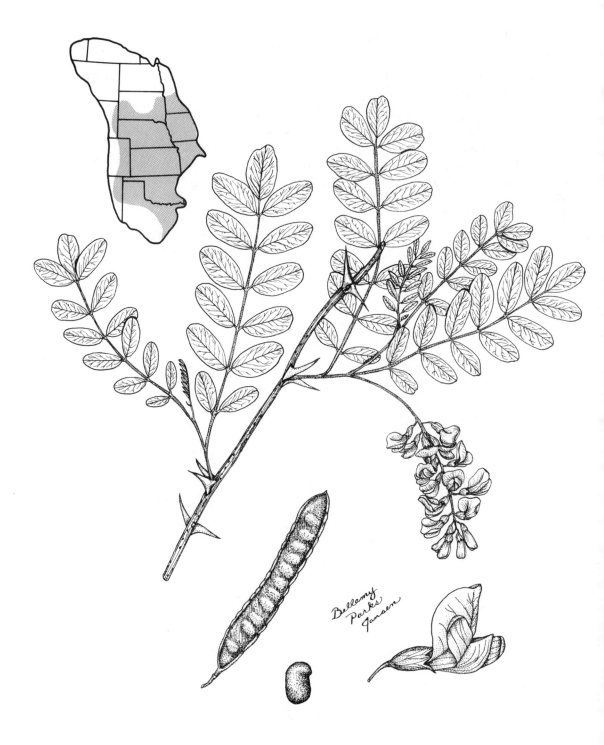

Robinia pseudoacacia L.   Black locust (Figure 97)

[*pseudes* (Gk.): false; + *akakia* (Gk.): acacia, in reference to the appearance of this species being similar to that of acacia.]

Life Span: perennial. *Origin*: native. *Height and Form*: tree to 15(18) m with a rounded or oblong crown. *Stems*: woody, 1.8–2.1 mm in diameter, gray or reddish-brown; spines scattered, short, in pairs, extending at right angles; trunks to 1.2 m in diameter; forming colonies from root sprouts. *Leaves*: deciduous, alternate, odd-pinnately compound, 1–2.5(3.5) dm long; leaflets 7–19(29), elliptic to oval, 2–3.5(5) cm long, 1–1.2 cm wide, apex mucronate, base obtuse, margin entire, upper surface glabrous, paler beneath and slightly pubescent; petiole 1–5 cm long, glabrous or pubescent, base enlarged; petiolules short; stipules linear-subulate, membranous at first, then developing into spines, 3–25 mm long. *Inflorescences*: axillary racemes, (0.5)1–1.5(2) dm long, with 10–40 flowers. *Flowers*: papilionaceous, fragrant; calyx tube campanulate, 4–5 mm long, sometimes purple, pubescent, 5-lobed; lobes bilabiate, lower 3 acuminate and 1–2 mm long, upper 2 connate and 2–3.5 mm long; corolla white with a small yellow patch on the banner; banner suborbicular, 1.5–2.5 cm long, claw 4–6 mm long; wings 2–2.2 cm long, lobed, claws 6–7 mm long; keel 1.9–2 cm long, semicircular, tip acute; stamens 10, diadelphous; pedicels 5–6(10) mm long, sometimes purple, pubescent. *Fruit*: legumes straight, flat, brown, 5–10 cm long, 1–1.5 cm wide, with (2)4–8(12) seeds, short-stipitate, tip apiculate, valves rigid; seeds hard, smooth, brown, mottled with darker brown or purple, to 5 mm long and 3 mm wide, slightly beaked. n=10.

Other Common Names: false acacia, post locust, shipmast locust, yellow locust

Black locust is common in moist but well drained soils in pastures, roadsides, valleys, and thickets. It has been planted over a wide area and has commonly escaped. It flowers in May and June. Domestic livestock and deer browse the foliage. Quail and squirrels eat the seeds. Bees utilize the nectar in making honey.

Cattle, horses, sheep, and humans have been poisoned by black locust. Cattle are poisoned after eating young shoots, and sheep are poisoned by the seeds. Horses have been poisoned by stripping and eating bark while being tied to black locust trees. Symptoms are anorexia, lassitude, weakness, nausea, and diarrhea. Poisoning is rarely fatal. Leaves, bark, roots, and seeds are poisonous.

Black locust was once commonly used for fence posts and railroad ties. Flowers can be eaten fresh, after being fried in a batter, or cooked as a vegetable. Flowers can also be used to make a tea.

Plants are usually available from commercial sources. Its popularity has decreased over the last century. It can become a serious weed in pastures. It has been used extensively for revegetating gullies and other eroded sites.

22. SOPHORA L.

[*sufayra* (Ar.): yellow, referring to color of a dye made from dry, immature flower buds.]

Unarmed deciduous or evergreen trees, shrubs, or perennial herbs; leaves odd-pinnately compound, with several to numerous leaflets; stipules minute, caducous; inflorescences of terminal or axillary racemes; flowers yellow, white, or purple; calyx 5-lobed; banner obtuse, tapered to the base; wings straight, clawed; keel nearly straight, clawed; stamens, distinct; legume more or less constricted between seeds, stipitate, woody or fleshy, terete, many-seeded, indehiscent.

A genus of about 70 species which are scattered in the warmer regions of the world. Three species grow in the Great Plains. Only one is relatively common, the second extends only into the southern Great Plains, and the third is an introduced ornamental.

Figure 98   *Sophora nuttalliana*

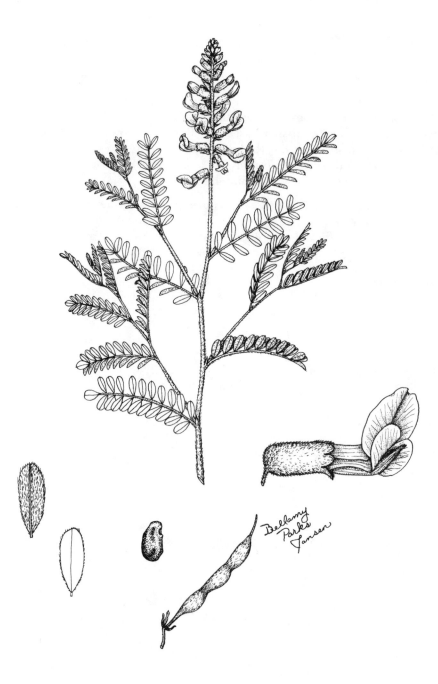

Sophora nuttalliana B. L. Turner

Silky sophora (Figure 98)

[*nuttalliana*: named for Thomas Nuttall (1787–1859), explorer and naturalist. He was first to collect this species.]

Life Span: perennial. *Origin*: native. *Height*: 1–4(7) dm. *Stems*: herbaceous, erect or ascending, branching above, branching or single from the base, arising from taproots and forming colonies by means of shoots from adventitious buds on lateral roots, silky-canescent. *Leaves*: alternate, odd-pinnately compound; leaflets 7–23(31), obovate or narrowly oblong to elliptic, 7–11(18) mm long, 1.5–6(9) mm wide, apex acute to obtuse (rarely retuse), base acute, silky-canescent; petioles short; stipules obsolete or bristlelike. *Inflorescences*: axillary or terminal racemes, lax to densely flowered, 7–11 cm long; peduncles 1–4 cm long; bracts 4–11 mm long, lanceolate, semipersistent. *Flowers*: calyx tube campanulate, gibbous, 5–7(8) mm long, strigulose, 5-lobed; lower 3 lobes 1–2.5 mm long, upper 2 lobes connivent for 1/2 of their length, short-acuminate to triangular, 1–2 mm long; corolla papilionaceous, white to ochroleucous (rarely purple); banner 1.2–1.6 cm long, strongly recurved, widened above the middle, claws 5–7 mm long; wings asymmetric, oblong, 8–11 mm long, claws 5–6 mm long; blades of keel 7–8 mm long, claws 5–6 mm long; 2 apical beaks 1–3 mm long; stamens 10. *Fruit*: legumes 2–5(7) cm long, short-stipitate, terete, beaked, constricted between the seeds, tardily indehiscent, with 1–4(7) seeds; seeds oblong, 4.5–5 mm long, 2.5–3 mm wide, olive brown, smooth.

Synonym: *Sophora sericea* Nutt.
Other Common Names: Nuttall's sophora, white loco

Silky sophora is infrequent to common in dry prairies, badlands, roadsides, abandoned fields, and stream valleys. It is most common on sandy soils. It has little forage value. It has been reported to be poisonous to horses, but toxicity was not proven in feeding trials. Alkaloids have been isolated from the seeds.

Silky sophora seed is not available from commercial sources. It is not valuable for erosion control or for landscaping.

Sophora affinis T. & G., Texas sophora, has leaves with 9–15 leaflets, rosy-white flowers, and black legumes. This woody plant forms thickets and grows in southern Oklahoma and north and central Texas. *Sophora japonica* L., Japanese pagodatree is an introduced ornamental. It has large leaves and clusters of flowers to 40 cm long. This small tree has a rounded crown and short trunk.

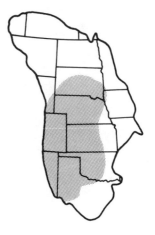

23. STROPHOSTYLES Ell.

[*strophe* (Gk.): turning; + *stylos* (Gk.): pillar, referring to the recurved style.]

Annual or perennial, twining or trailing herbs; leaves pinnately 3-foliate; inflorescences of axillary racemes, short, capitate, long-peduncled; flowers purplish-pink to white; calyx 4-lobed through fusion of upper 2 lobes; lowest lobe longest, exceeding the tube; banner orbicular to broadly obtuse-ovate; wings shorter than the keel, oblong, curved upward; stamens 10, diadelphous; legumes subterete, elongate, coiled after dehiscence.

Three species have been described. One perennial species is found in the southeastern United States. The other 2 are annuals and grow in the Great Plains.

A. Leaflets lanceolate to narrowly ovate, 2 times or more longer than wide; calyx tube densely hirsute; flowers 5–8 mm long; legumes 2–4 cm long ....2. *S. leiosperma*
A. Leaflets ovate to ovate-oblong, less than 2 times longer than wide; calyx tube glabrous to sparsely pubescent, flowers 8–14 mm long; legumes 4 cm long or longer ....1. *S. helvola*

Figure 99   *Strophostyles helvola*

1. *Strophostyles helvola* (L.) Ell.   Trailing wildbean (Figure 99)

[*helvolus* (Lat.): light-yellow, probably referring to the flowers.]

Life Span: annual. *Origin*: native. *Height*: twining vine, 3–25 dm long. *Stems*: herbaceous, twining, branched at the base, with taproots. *Leaves*: leaves on lower 1 to 4 nodes opposite, simple, 8–12 mm long; principal leaves alternate, pinnately 3-foliate; leaflets variable in shape, usually ovate-oblong or ovate (sometimes rhombic), 2–7 cm long, 1.5–5 cm wide, sparsely pilose on both sides or glabrous above, some with a lateral lobe, apex slightly rounded, generally mucronate; petioles 1–9 cm long; stipules, lanceolate, persistent, 4–6 mm long. *Inflorescences*: axillary heads of few to several flowers; peduncles 6–22 cm long; bracteoles lanceolate, acute, extending beyond the sinuses of the calyx. *Flowers*: calyx tube campanulate, glabrous or with a few appressed hairs, 1.5–3 mm long; upper lobe with 2 united teeth, lower lobe longest, 4–6 mm long; corolla purplish-rose to greenish-purple, fading to pinkish-yellow; banner 1–1.4 cm long, 1–1.2 cm wide, claw 1–1.5 mm long; wings 6–9 mm long, claw 1.5–2 mm long; keel 1.2–1.4 cm long, claw 2–3 mm long; stamens 10, diadelphous. *Fruit*: legumes, 4–9 cm long, 5–8 mm wide, subterete, sparsely appressed pubescent; seeds brownish-black, 5–10 mm long, reniform, sometimes scurfy, but with the scurfy coating readily detaching. n=11.

Synonym: *Strophostyles missouriensis* (S. Wats.) J. Small
Other Common Names: ambreique bean, tangle mealybean

Trailing wildbean is infrequent to locally common in prairie ravines, rocky woodlands, and open woods. It is also found along streams and on roadsides. It flowers from June to October. It is palatable to all classes of livestock. It is not highly valued, because it usually is not abundant. Its seeds are eaten by many kinds of birds and rodents.

Trailing wildbean is potentially valuable for erosion control in moist soils. Seed is not commercially available. It has no landscaping value.

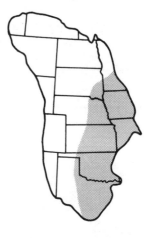

257

Figure 100   *Strophostyles leiosperma*

2. *Strophostyles leiosperma* (T. & G.) Piper   Smoothseed wildbean (Figure 100)

[*lius* (Gk.): smooth to the touch; + *sperma* (Gk.): seed, referring to the smooth seed.]

Life Span: annual. *Origin*: native. *Height*: twining vine, 2–10(20) dm long. *Stems*: herbaceous, twining, appressed-pubescent, with 1 to several branches at the base, with taproots. *Leaves*: leaves on lower 1–4 nodes opposite, simple, 8–10 mm long; principal leaves alternate, pinnately 3-foliate; leaflets narrowly ovate or lanceolate, 2–6 cm long, 2–18 mm wide, pilose on both sides, more densely so beneath, apex obtuse to acute, sometimes mucronate, margins entire. *Inflorescences*: axillary heads of 1 to few flowers; peduncles slender, (1)4–10 cm long; bracteoles lanceolate, hirsute. *Flowers*: upper lobe of calyx with 2 teeth, 1.5–2 mm long; middle tooth of lower lobe longest, 2–4 mm long, hirsute; corolla light rose to purple (rarely green) 5–8 mm long; banner 7–8 mm long, claw 1 mm long; wings 5–6 mm long, claw 1.5–2 mm long. *Fruit*: legumes, 2–4(5) cm long, subterete, elongate, with appressed or ascending pubescence (rarely glabrate); seeds gray to brown, sometimes mottled with purple or black, 2.5–4 mm long, 2–3 mm thick, round to reniform, with easily detached scurfy pubescence. n=11.

Synonyms: *Phaseolus leiosperma* (T. & G.) Piper, *Strophostyles pauciflora* (Benth.) S. Wats.
Other Common Names: littleflower mealybean, slickseed bean, small wildbean, wildbean vine

Smoothseed wildbean is scattered to locally abundant in dry or moist soils of prairies, woods, dunes, and shores. It flowers from May to October. It is not highly valued as forage, because it is generally not abundant. Foliage is eaten by deer. Birds and rodents eat the seeds.

Seed is not commercially available. It has limited value for erosion control in moist soils. It has no value as a landscaping plant.

259

24. STYLOSANTHES SW.

[*stylos* (Gk.): pillar or column; + *anthos* (Gk.): flower, in reference to the stalklike hypanthium and calyx tube.]

Herbaceous perennials with alternate, pinnately 3-foliate leaves; inflorescences of few-flowered, spikelike racemes; flowers small, usually yellow; calyx 4- or 5-lobed; lowest lobe longest; banner obcordate to broadly obovate or obtuse, longer than the other petals; narrowed at the base; wings oblong to obovate, clawed; keel curved upward, nearly as long as the banner; stamens 10, monadelphous, anthers alternately subglobose and oblong; fruit a loment, usually 2-jointed, lowest joint sterile.

This genus is represented by about 40 species found primarily in the tropics. They are most numerous in tropical America. One species occurs in the Great Plains.

Figure 101 *Stylosanthes biflora*

Stylosanthes biflora (L.) B.S.P.   Pencilflower (Figure 101)

[*bis* (Lat.): two; + *flos* (Lat.): flower, referring to the commonly 2-flowered inflorescences.]

Life Span: perennial. ***Origin***: native. ***Height***: 1.5–3.5(6) dm. ***Stems***: herbaceous, wiry, erect to spreading or ascending, more or less branched, glabrous to minutely pubescent or sparsely bristly, from a caudex. ***Leaves***: alternate, pinnately 3-foliate; leaflets elliptic or oblanceolate (lowermost sometimes obovate), 1–4 cm long, 2–7 mm wide, apex acute to obtuse, base acute, margins sparsely bristly, with conspicuous veins beneath; petiolule of terminal leaflet 1–3 mm long; petioles 1–3 mm long; stipules adnate to the base of the petiole, distinct portions linear or subulate and 3–8 mm long. ***Inflorescences***: racemes, from axils near the ends of branches, subcapitate or spikelike, 1- to 6-flowered, commonly 2-flowered. ***Flowers***: calyx tube campanulate above a slender hypanthium, 4 mm long, 5-lobed; lower and lateral lobes acute, nearly as long as or longer than the tube; upper 2 lobes rounded and united to near the apex; corolla orangish-yellow (drying pink or cream); banner 5–10 mm long, 5–6 mm wide, orbicular; wings 3–5 mm long, short-clawed; keel curved upwards, about as long as the wings; stamens 10, monadelphous, anthers alternating subglobose and oblong. ***Fruit***: loment, obliquely ovate, slightly flattened, 3–5 mm long, terminated by a short curved or coiled style, minutely pubescent, usually of 1 or 2 segments, proximal segment usually sterile; fertile seeds yellow to tan, slightly flattened, 2–2.5 mm long, 1.5 mm wide, often triangular. n=10.

Synonyms: *Stylosanthes biflora* var. *hispidissima* (Michx.) Pollard & Ball, *S. riparia* Kearn.

Pencilflower is scattered to locally common in dry, rocky or sandy soils of prairies, open woods, and roadsides. It is most common in acid soils. It flowers from May to September. Domestic livestock and wildlife consume the foliage. Rodents and birds eat the seeds.

Seed of pencilflower is not available from commercial sources. It has little value for landscaping or erosion control.

25. TEPHROSIA Pers.

[*tephros* (Gk.): ash-colored, referring to the gray pubescence.]

Herbaceous perennials, more or less woody at the base; leaves odd-pinnately compound; flowers yellow to purplish-rose, in terminal or lateral racemes; calyx tube hemispheric to campanulate, slightly oblique, 5-lobed; lobes lanceolate, exceeding the tube, upper lobes slightly shorter than lower lobes; petals all clawed; banner subround, short clawed, sericeous on back; wings and keel connivent, broadly auriculate on upper side above the short claw; wings ovate-oblong; keels semicircular; stamens 10, upper 1 at least partially free; legume linear, several-seeded.

About 250 species occur in tropical and subtropical regions in both hemispheres. Species are most abundant in Africa, several grow in the southern United States, and only 1 species is found in the Great Plains.

Figure 102   *Tephrosia virginiana*

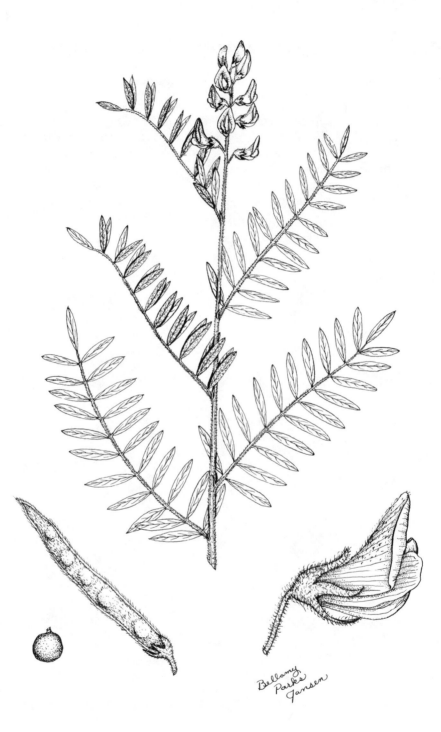

Tephrosia virginiana (L.) Pers. Goat's rue (Figure 102)

[*virginiana*: of or from Virginia.]

Life Span: perennial. *Origin*: native. *Height*: 2–7 dm. *Stems*: one to several from a woody caudex, ascending or erect, villous to glabrous. *Leaves*: alternate, odd-pinnately compound, 6–15 cm long, nearly sessile; leaflets 9–31, elliptic to linear oblong, 1–3 cm long, 4–10 mm wide, acute or obtuse, terminal leaflet mucronate, sparsely appressed pubescent to glabrous above, densely silky and gray beneath; petioles usually shorter than lowermost leaflets; stipules 8–10 mm long, subulate. *Inflorescences*: terminal racemes, simple (rarely branched), compact, 4–10 cm long, villous; peduncles short. *Flowers*: calyx tube campanulate, 3–5 mm long, densely strigose to villous, 5-lobed; lobes triangular or lanceolate-acuminate, longer than the tube; corolla papilionaceous, bicolored; banner 1.4–2.1 cm long, nearly as wide, lemon-yellow outside, cream inside, sericeous on back, claw 2–3 mm long; wings rose, 1.5–2 cm long, claw 2–3 mm long; keel rose, 1.4–1.5 cm long, claw 2–3 mm long; stamens 10, monadelphous; pedicels 4–18 mm long. *Fruit*: legume, 2–6 cm long, straight or slightly curved, flattened, sparsely to densely strigose, 5- to 11-seeded; seeds round to reniform, yellowish-brown to green, variegated with black, flattened, 3–5 mm long, white membrane readily flaking loose from smooth surface. n=11.

Synonyms: *Cracca latidens* J. Small, *C. leucosericea* Rydb., *C. virginiana* L., *Tephrosia leucosericea* (Rydb.) Cory
Other Common Names: catgut, devil's shoestring, Virginia tephrosia, wild sweetpea

Goat's rue is infrequent to locally abundant on dry prairies, open fields, open woods, and dunes. It is most frequent on sandy soils and on sites that are periodically burned. It flowers from May to July. Cattle rarely eat goat's rue, but it decreases on heavily grazed range. Deer consume the foliage, and upland birds and rodents utilize the seeds.

Seed is not commercially available. It has a limited potential for erosion control on sandy soils. It has no value for landscaping.

Goat's rue contains rotenone. Indians are reported to have used the plant to poison fish. They used the crushed roots as vermifuge.

26. THERMOPSIS R. Br.

[*thermos* (Gk.): lupine, + *opsis* (Gk.): appearance, in reference to the plant having an appearance similar to that of a lupine.]

Perennial herbs with 3-foliate leaves and foliaceous stipules; inflorescences of racemes; calyx bilabiate, upper lip shallowly lobed or merely notched, lower lip deeply 3-lobed; petals nearly equal in length, banner suborbicular, wings and keel oblong; stamens 10, distinct; ovary sessile or short-stipitate; legume linear, flat, several-seeded, sessile or short-stipitate.

About 20 species have been described in North America and eastern Asia. Only 1 species is common in the Great Plains.

Figure 103   *Thermopsis rhombifolia*

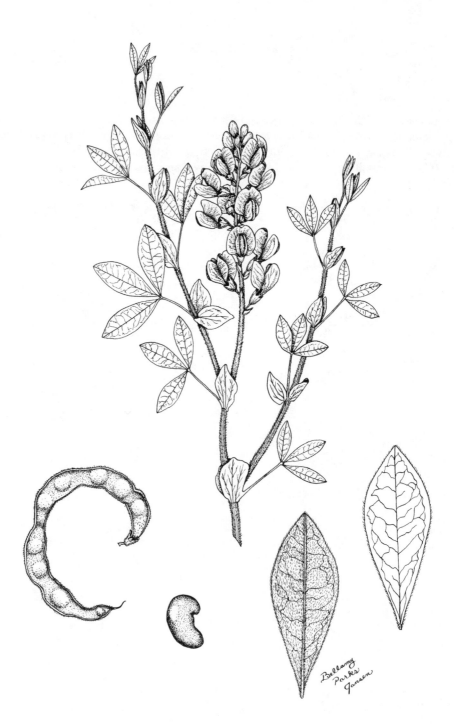

Thermopsis rhombifolia Nutt. *ex* Richards.　　　　　Goldenpea (Figure 103)

[*rhombos* (Gk.): 4-sided figure whose opposite sides and angles are equal; *folium* (Lat.): leaf, referring to the shape of the leaflets.]

Life Span: perennial. *Origin*: native. *Height*: 1–3(6) dm. *Stems*: herbaceous, erect or ascending from rhizomes, grooved, appressed-pubescent to glabrate. *Leaves*: alternate, palmately 3-foliate; leaflets obovate to elliptic, 2–4(5) cm long, 2–5 mm wide, glabrous (especially above) to appressed-pubescent (especially below); stipules foliaceous, broadly ovate to ovate-lanceolate. *Inflorescences*: subterminal racemes, up to 1 dm long, with 10 to 30 flowers. *Flowers*: calyx tube campanulate, pubescent, bilabiate, 4–5 mm long, upper lobe with 2 united teeth; corolla papilionaceous, petals yellow, nearly equal in length; banner suborbicular, may have purple dots; wings and keel oblong; stamens 10, distinct; pedicels 4–10 mm long. *Fruit*: linear, arcuate to curved in a half-circle or more, 6–10 cm long, flat, coriaceous, pubescent to glabrous at maturity; seeds green to black, smooth, reniform, 3–4(5.5) mm long. n=9.

Synonym: *Thermopsis arenosa* A. Nels.
Other Common Names: buffalo bean, false lupine, golden banner, golden bean, prairie buckbean, yellow bean

Goldenpea is infrequent to common on dry plains, hills, and slopes. It is often in sandy soil along roads, railroads, and in open woods. It blooms from April to July. Goldenpea has practically no forage value. It contains at least two poisonous quinolizidine alkaloids. Livestock losses are seldom reported because animals rarely eat the plants. Sheep have been nonfatally poisoned by the seeds. Elk and deer occasionally eat the plant. Circum-stantial evidence links goldenpea with poisoning of children.

Goldenpea generally has a high seed set. Commercial seed is seldom available. Seeds require scarification and/or stratification to improve germination. It can also be propagated by rhizome cuttings, and branch tips will root in mist benches. Its showy flowers and attractive foliage make it potentially useful in landscapes. It also has a potential value in erosion control because of its deep roots and rhizomes.

Lakota Indians used the flowers to treat inflammatory rheumatism. Flowers were mixed with hair and burned. Smoke was caught under a robe or blanket and inhaled.

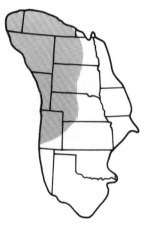

27. TRIFOLIUM L.

[*tres* (Lat.): three; + *folium* (Lat.): leaf, in reference to the compound leaves with three leaflets.]

Annual, biennial, or perennial herbs with weak stems, erect to decumbent or stoloniferous; leaves palmately or pinnately 3-foliate (rarely more), leaflets serrulate or denticulate, elliptic to oblong or ovate to obovate, pubescent or glabrous; petioles well developed; stipules adnate to the petioles; inflorescences of sessile or peduncled racemes, terminal or axillary, capitate or spicate; flowers papilionaceous, yellow, white, pink, rose, red, or purple; calyx campanulate to tubular, sometimes oblique, usually bilabiate, lobes triangular to setaceous, often unequal; petals separate or more or less united into a tube; banner ovate to oblong or obovate, folded about the wings in bud but reflexed in flower; wings more or less hooked over the keel; stamens 10, diadelphous; legumes short, membranous, straight, often enclosed by the persistent calyx and corolla, 1- to 6-seeded, indehiscent.

About 300 species have been described. Most are in the northern temperate zone. Thirteen occur in the Great Plains, but only 7 are common.

A. Flowers yellow . . . . .1. *T. campestre*
A. Flowers red, pink, white, or purple
  B. Flowers sessile or with pedicels 1 mm long or less
    C. Plants perennial
      D. Plants with creeping stems, rooting at the nodes; heads not subtended by a
        pair of leaves . . . . .2. *T. fragiferum*
      D. Plants erect to decumbent, without creeping stems; heads subtended by a
        pair of leaves . . . . .5. *T. pratense*
    C. Plants annual. . . . . .4. *T. incarnatum*
  B. Flowers with pedicels of 2 mm longer or more (especially in fruit)
    E. Plants with creeping stems, rooting at the nodes. . . . . .7. *T. repens*
    E. Plants erect to ascending, not creeping, not rooting at the nodes
      F. Calyx lobes more than 2 times longer than the tube, tube 10-nerved
      . . . . .6. *T. reflexum*
      F. Calyx lobes less than 2 times longer than the tube, tube 5-nerved
      . . . . .3. *T. hybridum*

Figure 104 *Trifolium campestre*

Bellamy
Parks
Jansen

1. *Trifolium campestre* Schreb.

Plains clover (Figure 104)

[*campestre* (Lat.): of fields, a reference to its habitat.]

Life Span: annual or winter annual. **Origin**: introduced (from Europe). **Height**: 1–4 dm. **Stems**: ascending to decumbent, glabrate to finely appressed-pubescent, much branched, with a taproot. **Leaves**: alternate, pinnately 3-foliate; leaflets obovate to oblanceolate or elliptic, base acute, apex obtuse or retuse, denticulate above the middle, glabrous to sparsely pubescent, 5–15 mm long, 3–8 mm wide; terminal petiolule longer (1–3 mm long) than the lateral ones (0.2–0.5 mm long); petioles 2–3 cm long below, only 1–2 mm long above; stipules ovate-lanceolate, 5–8 mm long, adnate to petiole for 1/2 their length. **Inflorescences**: axillary heads, globose to short-cylindric, 6–15 mm long, 6–10 mm wide, 20- to 40-flowered; peduncles longer than the subtending leaves. **Flowers**: calyx tube campanulate, 0.5–1 mm long, membranaceous, 5-nerved, 5-lobed; lobes long-acuminate or bristle-like, lower lobe 1–1.5 mm long; corolla yellow (drying light brown); banner 3.5–4.5 mm long, 2–4 mm wide, prominently veined, longer than the wing and keel; stamens 10, diadelphous; pedicels 0.2–0.8 mm long. **Fruit**: oblong legume, 3 mm long, exserted from the calyx tube, 1-seeded, stipe 1 mm long; seeds yellowish-brown, ellipsoid, slightly flattened, smooth, 1 mm long. n=7.

Synonym: *Trifolium procumbens* sensu auctt. non L.
Other Common Names: large hopclover, low hopclover

Plains clover has escaped from seedings to pastures, roadsides, lawns, and open woods. It is infrequent to locally common, and it flowers from May to September. It is palatable to all classes of livestock. Use by wildlife has not been documented.

Seed is commercially available. It has limited applications for erosion control. Plains clover does not have a potential for landscaping.

Trifolium dubium Sibth., small hopclover or yellow sucklingclover, is also an annual with yellow flowers. It usually only has 5–18 flowers per head, and its petioles are shorter than the leaflets. It grows in the southeastern Great Plains.

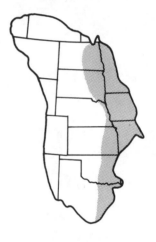

275

Figure 105   *Trifolium fragiferum*

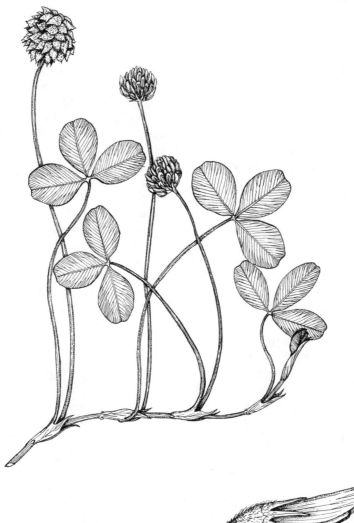

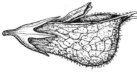

2. *Trifolium fragiferum* L.

Strawberry clover (Figure 105)

[*fragum* (Lat.): strawberry; + *fer* (Lat.): bearing, in reference to the strawberry-like appearance of the inflorescences.]

Life Span: perennial. *Origin*: introduced (from Europe). *Height*: 1–4 dm. *Stems*: several stems from taproots, herbaceous, pubescent, creeping and rooting at the nodes. *Leaves*: alternate, palmately 3-foliate; leaflets broadly elliptical to obovate, apex retuse, serrulate, 1–2.5 cm long; petioles 3–15(20) cm long; stipules lanceolate-subulate, 1.5–2 cm long, adnate to the petioles. *Inflorescences*: globose to cylindrical heads, pointed on top, 1–1.5 cm wide at anthesis, 1.5–3 cm wide in fruit; subtended by whorled bracts, 3–5 mm long, lowest bracts united to form an involucre; peduncles 4–20 cm long, longer than the leaves, usually pubescent. *Flowers*: calyx 3.5–4.5 mm long at anthesis, reticulate-veined, bilabiate, upper lip inflated in fruit, lobes equaling tube in fruit; corolla pink to pinkish-white (rarely red), persistent; banner 6–8 mm long; stamens 10, diadelphous. *Fruit*: inflated legumes, 1- to 3-seeded; seeds green to dark brown, dull, smooth. n=8.

Strawberry clover is infrequently planted on wet saline and alkaline soils in the western Great Plains. It is excellent forage for all classes of livestock. It produces palatable and nutritious hay. There is no record of use by wildlife. It flowers from May to September.

Commercial seed is available. Scarification improves germination. Strawberry clover has been used as a green manure crop and for reclamation of moist, alkaline soils. It is also valuable under irrigation when drainage is a problem. It will withstand flooding for extended periods.

277

Figure 106 *Trifolium hybridum*

3. *Trifolium hybridum* L.   Alsike clover (Figure 106)

[*hybrida* (Lat.): mongrel or crossbred, because it was once thought to be a cross between 2 species of clover.]

Life Span: perennial. *Origin*: introduced (from Europe). *Height*: 2–8 dm. *Stems*: herbaceous, numerous from taproots, erect to ascending, somewhat cespitose, glabrous to sparsely pubescent. *Leaves*: alternate, palmately 3-foliate; leaflets oval to elliptic, 1.5–3.5 cm long, 8–25 mm wide, apex obtuse to retuse, denticulate above, serrulate below, glabrous; petiolules 1 mm long; petioles slender, glabrous, 5–15 cm long below, only 1–3 cm long above; stipules ovate-lanceolate, tapering to an attenuate tip, adnate to the petiole for up to 1/2 of their length, 1–3 cm long, 3-veined. *Inflorescences*: axillary heads from upper axils, globose, numerous, 1.5–2.5(3.5) cm in diameter, not involucrate, 30- to 50-flowered; peduncles 2–8(12) cm long, equaling or exceeding the leaves. *Flowers*: calyx tube campanulate, glabrous, 1.5–2.5 mm long, 5-nerved; lobes linear-subulate, slightly unequal, longest equaling or exceeding the tube; corolla white to pinkish-white (drying brown); banner 5–11 mm long, 2–3 mm longer than the wings and keel, obovate-oblong; stamens 10, diadelphous; pedicels 1–2 mm long at anthesis, expanding to 5–7 mm long. *Fruit*: legume, linear-oblong, thin-walled, 3–4 mm long, exserted from the calyx tube, indehiscent, 2- to 4-seeded; seeds light green to greenish-black, cordate, smooth, 1.5 mm long. n=8.

Other Common Name: Swedish clover

Alsike clover has escaped from plantings to roadsides, waste grounds, fields, pastures, lawns, and stream valleys. It grows well on poorly drained acid soils. It flowers from May to October. It is good forage for all classes of livestock. Large quantities in diets have been reported to cause photosensitivity in horses, cattle, and sheep. The condition is called trifoliosis and is most prevalent in horses. Deer, pronghorn, and rabbits eat the foliage, and birds consume the seeds.

Commercial seed is available. It is a good plant for meadows, especially if the soils are acid. Alsike clover has no potential for landscaping. It is a good honey plant.

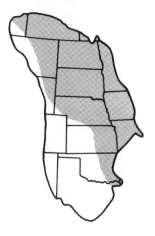

Figure 107   *Trifolium incarnatum*

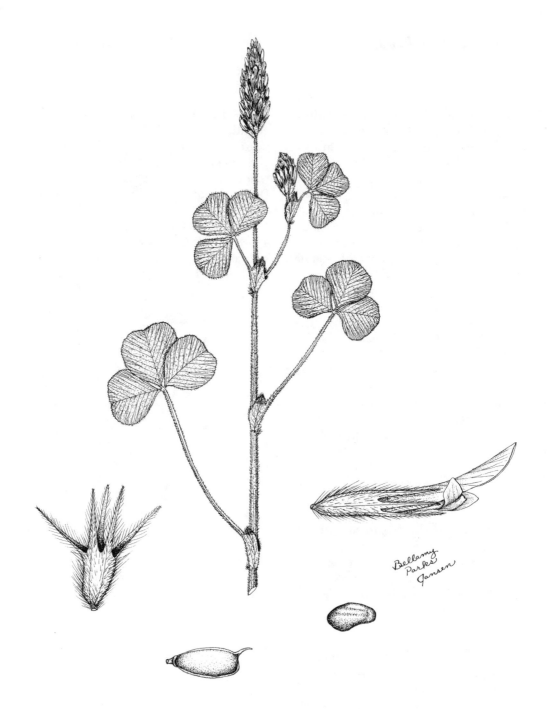

4. *Trifolium incarnatum* L.

[*incarnatus* (Lat.): made into flesh, in reference to the flower color.]

Life Span: winter annual. **Origin**: introduced (from southern Europe). **Height**: 2–8 dm. **Stems**: herbaceous, erect, appressed-pubescent, branching or rarely simple from a taproot. **Leaves**: alternate, palmately 3-foliate, pubescent; leaflets broadly obovate, 1–4 cm long, nearly as wide, tapering to the base, rounded or truncate at the summit, upper half erose to denticulate; lower petioles 5–20 cm long, upper petioles 5–30 mm long, densely pubescent; stipules erose, red to purple around the summit, 1–2 cm long, mostly adnate to the petioles. **Inflorescences**: terminal heads, solitary, ovoid to cylindric, 2–7 cm long, 1–2.5 cm in diameter, not involucrate, 75- to 125-flowered; peduncles erect, 4–12 cm long. **Flowers**: calyx tube cylindrical to campanulate, 3 mm long, densely villous, 10-nerved; lobes about equal, linear-subulate, equaling or longer than the tube; corolla crimson (rarely white), 8–13 mm long; banner oblanceolate, obtuse, 2–5 mm longer than the wing and keel petals, sessile; stamens 10, diadelphous. **Fruit**: legume, sessile, ovoid, enclosed in the calyx tube, 1-seeded; seeds ovoid, yellow or reddish-brown, shiny, smooth, 2–2.5 mm long. n=7.

Crimson clover is planted for pasture and hay. It has escaped locally to old fields, waste grounds, and roadsides. It flowers from May to August. It is nutritious and palatable to all classes of livestock. There is no record of use by wildlife other than birds eating the seeds. Hay from mature crimson clover may be dangerous to horses. The stiff, wirey hairs become felt-like balls that may block the intestines.

Seed of crimson clover is commercially available. It has been used for winter soil protection and as a green manure crop. It is a heavy user of phosphate. Its landscaping potential has not been explored.

Trifolium arvense L., rabbitfoot clover, is an annual species with a glabrous or uniformly pubescent calyx tube. Its flowers are rose to pink or white. It is found in a few scattered locations in the eastern Great Plains. *Trifolium resupinatum* L., Persian clover, is another annual species with pink to purple flowers. The lower lip of its calyx tube is glabrous and the upper lip is densely pubescent. It is rare in scattered locations. It is occasionally a contaminant in lawn seed.

Figure 108   *Trifolium pratense*

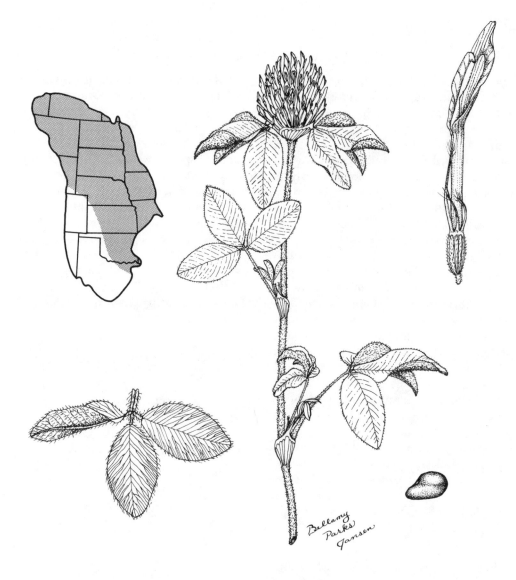

Bellamy
Parks
Jansen

5. *Trifolium pratense* L.

Red clover (Figure 108)

[*pratensis* (Lat.): growing in a meadow, in reference to its habitat.]

Life Span: biennial or short-lived perennial. **Origin**: introduced (from southern Europe). **Height**: 2–8(10) dm. **Stems**: several from a long taproot, herbaceous, erect to decumbent, cespitose, sparsely to densely appressed-pubescent, simple or short-branched. **Leaves**: alternate, palmately 3-foliate; leaflets ovate or broadly elliptic, 2–5 cm long, 1–3.5 cm wide, marked above with a light green crescent or inverted "v" (fading on drying), pubescent on both surfaces, apex obtuse (rarely retuse), base broadly acute, margins finely serrate on youngest leaflets, older leaflets entire; petiolules 1–1.5 mm; lower leaves with long petioles, upper leaves sessile or with short petioles; stipules membranous, oblong, distinct portion abruptly narrowed to a short awn, conspicuously veined, 1–3 cm long, adnate to the petiole for 1/2 to 3/4 of the length. **Inflorescences**: mostly terminal heads, 30- to 90-flowered, globose or round-ovoid, 1.5–2.5 cm long, 2–2.5 cm wide, subtended by a pair of reduced leaves, sessile or on peduncles to 2 cm long. **Flowers**: calyx tube campanulate, 3–4 mm long, membranous, sparsely pilose to glabrous, 10-nerved; lobes 5, setaceous, longest one 4–8 mm long, others 2–5 mm long; corolla rose to purplish-pink (rarely white), 1.2–2 cm long; banner obovate-oblong, equaling or slightly exceeding the wings, blade enveloping the other petals; stamens 10, diadelphous. **Fruit**: legume, ovoid to oblong, thickened above, 3 mm long, 2 mm wide, 1-seeded (rarely 2-seeded); seeds yellowish-brown to green mottled with purple, smooth, ellipsoid with a slight lateral lobe, 1.5–2 mm long. n=7, 14.

Synonym: *Trifolium medium* L.

Other Common Names: medium redclover, mammoth redclover

Red clover has spread from seeded fields to lawns, roadsides, waste ground, and stream valleys. It grows best on heavy, fertile, and well-drained soils. It flowers from May to September. It generally produces hay of high quality that is palatable to all classes of livestock. It is grazed by some species of wildlife and furnishes food and cover. It is an excellent honey plant.

Second cutting or late season hay may produce a syndrome in cattle, horses, and sheep characterized by slobbering. It may progress to bloating, stiffness of gait, decreased milk flow, and diarrhea. Symptoms stop soon after the animals consume a different hay. Red clover has been reported to rarely cause photosensitization. Dried flowers were used in past centuries in Europe to treat whooping cough and ulcers.

Red clover has little value for landscaping. Seed of several cultivars is commercially available. It is sometimes seeded in mixtures of other legumes and grasses. Scarification improves germination.

Trifolium beckwithii Brew. *ex* S. Wats., Beckwith's clover, is also a perennial with reddish to light purple flower. It is glabrous and in the Great Plains is found only in a few counties in South Dakota.

Figure 109　*Trifolium reflexum*

6. *Trifolium reflexum* L.

Buffalo clover (Figure 109)

[flexus (Lat.): bend, in reference to the recurved pedicels.]

Life Span: annual (or biennial). *Origin*: native. *Height*: 2–5 dm. *Stems*: 1 to several from taproots, herbaceous, ascending, glabrous to finely villous, branched from the base. *Leaves*: alternate, palmately 3-foliate, pubescent to glabrate; leaflets obovate to oblanceolate or broadly elliptic, 1–4 cm long, 1–1.5 cm wide, apex obtuse or occasionally retuse, base acute or short acuminate, margins finely serrate; petioles 0.5–14 cm long; stipules large, 1–3 cm long, foliaceous, lanceolate to ovate, adnate to the petiole for 1/4 their length. *Inflorescences*: terminal and axillary heads, globose, 2–3 cm across at anthesis; peduncles 2–10 cm long. *Flowers*: calyx tube 1.2–1.7 mm long, membranaceous, strongly 10-nerved, glabrous to short pubescent; lobes linear or long-acuminate, 4–7 mm long; banner pink to reddish-pink, 8–12 mm long, 5–6 mm wide, slightly exceeding the keel and wing petals; wings and keel pink or white; stamens 10, diadelphous; pedicels slender, to 1 cm in fruit, recurving. *Fruit*: oblong legume, 2- to 4-seeded, body 3–5 mm long; stipe 1–1.5 mm long; seeds yellowish-brown, 1–1.5 mm long, 1 mm wide. n=8.

Buffalo clover is scattered and not common in upland prairies, open woods, and gravelly stream valleys. It usually grows in acid soils. It flowers from May to August. It is palatable and nutritious to domestic livestock and wildlife. It is not considered to be an important constituent of their diets because of its lack of abundance.

Seed is not commercially available. Buffalo clover has limited value for erosion control. It has little potential for landscaping.

Trifolium carolinianum Michx., Carolina clover, is similar in appearance, except that it has yellowish-white or purple flowers. The heads are only 1–1.5 cm in diameter. It grows in the southeastern Great Plains.

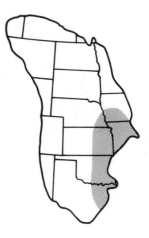

Figure 110   *Trifolium repens*

7. *Trifolium repens* L. White clover (Figure 110)

[*repens* (Lat.): creeping, in reference to the stems.]

Life Span: perennial. **Origin**: introduced (from Europe). **Height**: (0.5)1–4 dm. **Stems**: herbaceous, glabrous to slightly pubescent, creeping, rooting at the nodes, mat-forming, from taproots. **Leaves**: palmately 3-foliate, glabrous; leaflets 1–3 cm long, 5–20 mm wide, broadly elliptic to obovate, obtuse to retuse at the summit, base acute, margins serrulate to denticulate; petioles 3–20 cm long; base of stipules united into a membranaceous sheath, 3–10 mm long. **Inflorescences**: axillary heads without involucres, globose, 1–3 cm across, 20- to 90-flowered; peduncles 5–20 cm long, longer than the leaves. **Flowers**: calyx glabrous, tube 2–4 mm long, cylindrical, 10-nerved, 5-lobed; lobes narrowly triangular, acuminate, 1–2 mm long, unequal; corolla white or tinged with pink, 7–12 mm long; banner elliptic-obovate, obtuse at the summit, exceeding the obtuse wings; stamens 10, diadelphous; pedicels 1–2 mm long, elongating to 5–7 mm in fruit. **Fruit**: legume, oblong to linear, 4–6 mm long, 2- to 4-seeded; seeds cordate, yellow, smooth, 1–1.4 mm long. n=16, 24, 32.

Synonym: *Trifolium saxicola* J. Small
Other Common Names: ladino clover, wild whiteclover, honeysuckle clover, Dutch clover

White clover is common in lawns, pastures, fields, stream valleys, roadsides, and waste places. It grows in all soil textures, but it is most common in silt and clay. It was brought to North America by some of the first settlers. It flowers from May to October. It has excellent forage quality and palatability to all classes of livestock, but yields are generally low in the Great Plains. Deer, elk, and pronghorn eat the foliage, and numerous birds and rodents eat the seeds.

Bloat may occur in cattle grazing lush white clover. It may contain cyanogenic glycosides, but no loss of life has been attributed to hydrocyanic poisoning.

White clover is valuable for hay, pasture, erosion control, and soil improvement. It is also an excellent honey plant, and insects are necessary for pollination. Seeds of several cultivars are available. Proper use of fertilizers increases yields.

Trifolium stoloniferum Muhl. *ex* Eat., running buffaloclover, is also a creeping plant that roots at the nodes. Calyx lobes are generally more than 2 times as long as the tube. Corolla is white and tinged with purple. It is rare in the Great Plains and can only be found in the eastern portion.

28. VICIA L.

[*vicia* (Lat.): ancient name of vetch.]

Annual, winter annual, biennial, or short-lived perennial herbs; stems erect or more commonly climbing or trailing; leaves even-pinnately compound, with 2–24 leaflets; terminal leaflets in most species modified into tendrils; petioles obsolete or to 3 mm long; stipules small, herbaceous, persistent, often semi-sagittate; inflorescences of few-flowered axillary clusters; calyx regular or irregular, often gibbous at the base, 5-lobed; corolla papilionaceous and yellow, white, purple, or blue; banner apically notched, with a broad claw overlapping the wings, blade obovate to subround; wings oblong or narrowly obovate, adherent to the keel, usually exceeding the keel; stamens 10, diadelphous; ovary sessile or short-stipitate; style filiform, pubescent only at the summit; legumes flat to terete, dehiscent, 2-valved, sometimes transversely separate between the seeds, 2-several seeds.

A genus of over 150 species distributed on all continents. Fifteen are native mainly to the southern and eastern portions of the United States. Eight occur in the Great Plains, and 2 are common.

A. Stipules with a single hastate basal lobe, densely villous; pedicel appearing laterally inserted . . . . .2. *V. villosa*
A. Stipules serrate, glabrous to sparsely pilose or strigulose-villous; pedicel attached basally or nearly so . . . . .1. *V. americana*

Figure 111   *Vicia americana*

1. *Vicia americana* Muhl. *ex* Willd.

American vetch (Figure 111)

[*americana*: of or from America.]

Life Span: perennial. **Origin**: native. **Height**: 2–11 dm. **Stems**: herbaceous, glabrous to sparsely pilose, sprawling to climbing, usually not vining. **Leaves**: alternate, even-pinnately compound, terminating in branched or simple tendrils; leaflets 4–16, narrowly oblong or elliptic to linear, 1–4 cm long, 1–9 mm wide, apex obtuse or truncate, mucronate, coriaceous or stiff with conspicuous nonbranched veins, entire (rarely toothed); stipules usually serrate. **Inflorescences**: axillary racemes on peduncles, usually shorter than subtending leaves, 2- to 10-flowered. **Flowers**: calyx tube 3.5–6 mm long, slightly gibbous at base, 5-lobed; lobes variable, unequal, lowest lobe lance-attenuate, 1–4 mm long, upper lobe short and broad; corolla bluish-purple (rarely white); banner 1.2–2.5 cm long, longer than the wings and keel. **Fruit**: legumes, 2.5–4 cm long, 2-valved, glabrous, 2- to 12-seeded; seeds dark brown, slightly flattened. n=7.

Two varieties occur in the Great Plains. Var. *americana* has vining stems 4–11 dm long, tendrils with several branches, leaflets over 4 mm wide, and 5–10 flowers in each raceme. Var. *minor* Hook. stems are usually not vining and less than 4 dm long. Its leaflets are usually less than 4 mm wide, and the racemes are 2- to 4-flowered.

Synonyms: *Vicia dissitifolia* (Nutt.) Rydb., *V. oregana* Nutt., *V. sparsifolia* Nutt., *V. trifida* Dietr.
Other Common Names: stiffleaf vetch, tasusu (Lakota)

American vetch is infrequent to locally common in upland prairies, badlands, roadsides, bluffs, and waste places. It flowers from May to August. It is excellent forage and is palatable to all classes of livestock. It quickly disappears with continued heavy use. It infrequently has been linked to cases of photosensitization. Deer, pronghorn, and rabbits eat the foliage, and numerous birds and rodents consume the seed.

Seed is not available commercially. American vetch has little potential for landscaping. Lakota indians boiled and ate young shoots and seeds.

Vicia cracca L., bird vetch, is a perennial that has been infrequently collected in the northern Great Plains. Its bluish-purple flowers are only 9–12 mm long, and the stipules are entire. *Vicia sativa* L., common vetch, is an annual or winter annual that is occasionally planted in the southern Great Plains. Its peduncles are much reduced or wanting, and it has 1–3 bluish-purple flowers per axil.

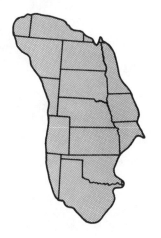

Figure 112   *Vicia villosa*

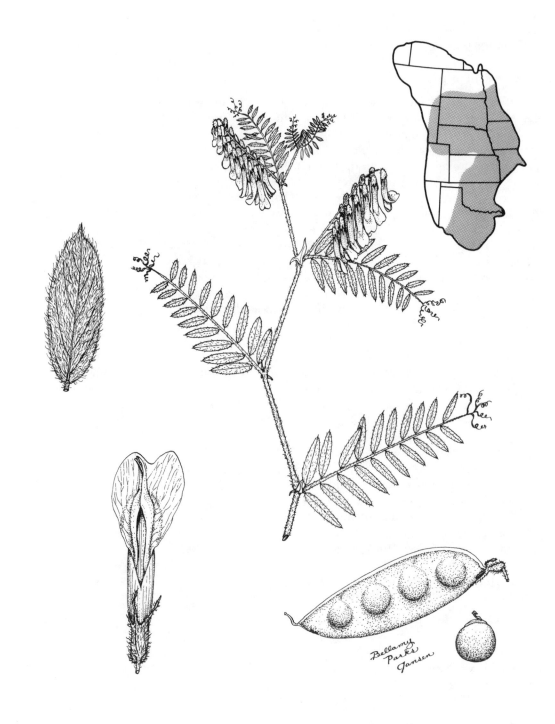

2. *Vicia villosa* Roth   Hairy vetch (Figure 112)

[*villosus* (Lat.): hairy, in reference to the stems, leaves, and peduncles.]

Life Span: most commonly annual, but sometimes biennial or even perennial. *Origin*: introduced (from Europe). *Height*: vines 5–20 dm long. *Stems*: herbaceous, villous, twining, climbing, from taproots. *Leaves*: alternate, even-pinnately compound, 6–15 cm long; leaflets 10–24, opposite or alternate along the rachis, narrowly oblong to linear-lanceolate, 1–2.5 cm long, 3–7 mm wide, apex obtuse and mucronate, softly villous with hairs 1–2 mm long; tendrils much branched; stipules lanceolate to ovate, 5–12 mm long, each with a single hastate basal lobe. *Inflorescences*: axillary racemes, dense, secund, 10- to 60-flowered, the long peduncle softly villous with hairs 1–2 mm long. *Flowers*: calyx irregular, villous, tube 2–4 mm long, gibbous at the base; upper lobes linear-triangular, 0.8–1.5 mm long; lateral and lower lobes linear above a triangular base, 1–5 mm long; corolla bluish-purple (rarely white), drying blue, 9–16 mm long, spreading blade of the banner oblong, 7 mm wide, less than 1/2 as long as the claw; pedicel apparently laterally inserted. *Fruit*: legume, oblong, flattened, 2–3 cm long, 7–10 mm wide, glabrous to pubescent, 3- to 5-seeded; seeds dark brown, globose, 4–5 mm in diameter. n=7.

Two varieties are recognized in the Great Plains. Var. *villosa* is the most common. Var. *glabrescens* Koch has a raceme with appressed or incurved hairs. It has also been known as *V. dasycarpa* Ten.

Other Common Names: winter vetch, woollypod vetch

Hairy vetch is planted in fields and along roads and occasionally escapes to other fields, waste places, and stream valleys. It is most common in sandy soils. It flowers from April to August. It is planted for hay or pasture and is usually sown with a support crop of small grain. Hay quality is excellent to good. Numerous species of wildlife utilize the foliage and seeds.

Several cultivars of hairy vetch are commercially available. It resists more alkali in the soil than do most legumes. It is also cold and drought hardy. It can be used for winter soil cover and as a green manure crop. It has no value for landscaping.

Relatively large quantities of seed may poison livestock. Another condition, hairy vetch poisoning, has been linked in the south to consumption of herbage. Its symptoms include dermatitis, conjunctivitis, and diarrhea. Mortality has been reported to be up to 50% of the affected animals.

Vicia exigua T. & G., little vetch, *V. leavenworthii* T. & G., Leavenworth's vetch, and *V. ludoviciana* Nutt., deer peavetch, are all winter annuals of the southern Great Plains. They are highly variable and difficult to separate. McGregor, in the *Flora of the Great Plains*, suggests treating *V. exigua* and *V. leavenworthii* as varieties of *V. ludoviciana*.

GLOSSARY

A- A prefix meaning without

Acaulescent Stemless, without an above-ground stem or apparently so

Acuminate Gradually tapering to a sharp point; compare with acute

Acute Sharp-pointed, but less tapering than acuminate

Adnate Attached or grown together; fusion of unlike parts

Adventitious Arising from a position not considered typical, as adventitious roots arising from the sides of a stem

Alternate Located singly at each node

Annual Within 1 year; applied to plants which do not live more than 1 year

Anther Pollen-bearing portion of a stamen

Anthesis Time when pollination occurs

Antrorse Directed upwards or forwards; opposite of retrorse

Apetalous Without petals

Apex The tip

Apical At the apex

Apiculate Abruptly constricted at the apex into a short, flexible point

Appressed Lying against an organ; flatly pressed against

Arcuate Curved or arched like a bow

Armed Provided with spines or thorns

Aromatic Having a fragrant odor

Articulate Jointed; provided with nodes

Ascending Growing upward

Attenuate Gradually narrowed to a slender apex or base

Auricle An earlike appendage

Awn A bristlelike appendage at the end, back, or edge of an organ

Axil Angle between two organs

Axillary Growing in an axil

Axis The portion of the stem within the inflorescence on which the flowers or fruits are borne

Banner Upper petal (standard) of the papilionaceous flower

Basifixed Attached by the base

Beak A hard point or projection

Bi- A prefix meaning 2

Biennial Living for 2 years

Bifid 2-lobed

Bilabiate 2-lipped, especially relating to the calyx and corolla

Bilocular Having 2 cavities

Bipinnate Twice pinnately compound

Blade The expanded portion of a petal or leaf

Bract A reduced leaf

Bracteate Having bracts

Bracteole A bract borne on a secondary axis, such as on a pedicel

Bristle A stiff, slender appendage

Bulb An underground bud with fleshy, thick scales

Caducous Early deciduous; falling off early

Calyx The sepals of a flower considered collectively, usually green bracts

Campanulate Shaped like a bell

Canescent Gray or white colored because of dense, fine pubescence

Capillary Fine and slender; hairlike

Capitate Headlike; arranged in very dense, headlike clusters

Caudate Bearing a slender taillike appendage

Caudex A short, usually woody, vertical stem located just below the soil surface

Caulescent Having an obvious stem

Cauline Pertaining to, or belonging to, the stem

Cespitose Growing in dense tufts or mats

Chasmogamous A flower that is open during pollination

Ciliate Fringed with hair on the margins

Claw The long, narrow base of a petal

Cleistogamous A flower, usually small, that remains closed and is self-pollinated

Coherent One organ sticking to another

Coiled Having a series of spirals

Compound Made up of 2 or more parts

Connate Fusion of like parts, such as petals to form a corolla tube

Connivent Coming together usually of like parts, but not securely fused

Conspicuous Obvious; easy to notice

Contiguous Adjacent or touching, but not fused

Cordate Heart-shaped

Coriaceous With the texture of leather

Corolla All of the petals considered collectively

Crenate Having rounded teeth; scalloped margins

Crenulate Finely crenate

Crown Persistent portion of the stem at the surface of the ground

Cuneate Wedge-shaped or triangular, with the narrow end at the point of attachment

Cuspidate Apex abruptly constricted into an elongated, sharp, and firm point

Deciduous Not persistent, but falling away at the end of the growing season

Declined Bent downward or forward

Decumbent Curved upward from a horizontal or inclined base

Decurrent Extending downward from the point of attachment

Decurved Curved downward

Deflexed Bent downward, usually at a sharp angle

Dehiscence Method of opening to disperse seeds (sometimes pollen)

Dehiscent Open regularly by slits or valves

Deltoid Triangular, shaped like the Greek letter delta

Dentate With pointed, coarse teeth spreading at right angles to the margin

Denticulate Minutely dentate

Diadelphous Describing stamens united by their filaments into 2 groups, often unequal in number

Diffuse Open; widely spreading

Dimorphic Having 2 forms

Distal Away from the place of attachment

Divaricate Widely and stiffly divergent

Diverging Widely spreading, less so than divaricate

Dolabriform Descriptive of hairs attached in the middle; malpighiaceous

Dorsal Relating to the back of an organ; opposite of ventral

Elliptic, ellipsoid Oval, pointed at the ends and widest in the middle

Elongate Narrow, the length many times the width or thickness

Emarginate Having a shallow notch at the tip

Endosperm Nutritive tissue, often containing starch, stored around the embryo of the seed

Entire Whole; with a continuous margin

Erose Irregularly notched at the apex; appearing gnawed or eroded

Exserted Protruding or projecting beyond; not included

Falcate Sickle-shaped

Fascicle A small bundle or cluster

Filament The stalk of the stamen which bears the anther

Filiform Threadlike, long and very slender

Fissure Narrow crack

Flexuous Bent alternately in opposite directions; a wavy form

Foliage Leaves considered collectively

Foliate Having leaflets

Foliose Leafy; bearing numerous leaves

Funnelform Shaped like a funnel

Fusiform Spindle-shaped; wide in the middle and tapering to both ends

Gamopetalous Having the petals at least partially united

Gamosepalous Having the sepals at least partially united

Gibbous Swollen on one side, usually at the base

Glabrate Nearly glabrous or becoming so with age; glabrescent

Glabrescent Nearly glabrous or becoming so with age; glabrate

Glabrous Without hairs; smooth

Glandular Supplied with glands

Glaucous Covered with a waxy coating that is easily rubbed off

Globose Nearly spherical in shape

Habit The characteristic form of a plant

Hastate Shaped like an arrowhead; but with the narrow, pointed basal lobes projecting perpendicularly to the midvein

Head A dense cluster of sessile or nearly sessile flowers

Hemi- A prefix meaning half

Herb A plant lacking woody, persistent aboveground parts

Herbage The green vegetative parts of a plant

Herbaceous Having the character of a herb; not woody, dying each year or dying back to the crown

Hirsute With straight, rather stiff, hairs

Hoary Grayish-white because of a dense, fine pubescence

Humistrate Lying flat on the soil surface

Hyaline Thin and translucent or transparent

Hypanthium A ring or cup around the ovary formed by a fusion of the bases of sepals, petals, and stamens; a modified receptacle

Imbricate Overlapping, like shingles on a roof

Included Not protruding or exserted

Incurved Bent inward

Indehiscent Not opening at maturity

Inflated Expanded as if by gas or air; puffed up; bladderlike

Inflorescence The flowering part of the plant; the mode of arrangement of flowers

Inserted Attached to or growing out of

Internode The part of the stem between 2 successive nodes

Introduced Not native, not an original member of the flora of North America

Intruded Thrust into another part

Involucre A whorl or circle of bracts below the flower or flower cluster

Involute Rolled inward from the edges, the upper surface within

Irregular Showing inequality in similar parts; asymmetrical

Keel The 2 united lower petals of a papilionaceous flower

Lanceolate Lance-shaped; rather narrow, tapering to both ends, widest below the middle

Lateral Belonging to or borne on the side

Lax Loose

Leaflet The subdivision of a compound leaf

Legume A usually dehiscent fruit composed of a single carpel having 2 sutures and with seeds attached along the ventral suture

Lenticel A corky spot in the bark or epidermis

Lenticular Lens-shaped; flattened, convex on both sides

Linear Long and narrow with parallel sides, or nearly so

Lobe A partial division of a calyx, leaf, or other organ

Locular Having locules

Locule A cavity or compartment of an organ, especially of an ovary

Loment A jointed fruit, constricted and breaking apart between the seeds

Lustrous Having a sheen

Maculate Blotched or mottled

Malpighiaceous Descriptive of hairs attached in the middle; dolabriform

Membranaceous, membranous Thin and somewhat transparent or translucent; like a membrane

Monadelphous Describing stamens with their filaments united in one group

Mucro A short, sharp, and slender point

Mucronate Terminated by a mucro

Multicipital Having many heads

Mycorrhiza A symbiotic relationship between roots of plants and certain fungi

Native Occurring in North America before settlement by European man

Nectary A gland or tissue which secretes nectar

Nerve Vascular bundles or simple veins of plant organs

Node Points along the stem where leaves are borne

Nodule A small knoblike enlargement; enlargements on the roots where rhizobial bacteria are active

Ob- A prefix meaning inverted; inversely

Obconic Cone-shaped with the point of attachment at the apex

Oblanceolate Lanceolate with broadest part above the middle and tapering toward the base

Oblique With unequal sides; asymmetrical

Oblong Longer than broad, with sides nearly equal

Obovate, obovoid Egg-shaped with the broadest part toward the top

Obsolete Not apparent; missing; rudimentary

Obtuse Rounded; blunt; with an angle greater than 90°

Ochroleucous Yellowish-white

Olivaceous Olive-green

Opposite Describing leaves arranged with 2 at each node and on opposite sides of the stem

Orbicular Nearly circular in outline

Oval Broadly elliptic

Ovary That portion of the pistil containing the ovules

Ovate, ovoid Egg-shaped with the broadest part near the base

Ovule The structure within the ovary that becomes the seed after fertilization

Palmate Divided in a handlike manner; leaflets or other plant parts arising from a common point

Panicle Compound inflorescence with a main axis and rebranched branches

Paniculate Resembling a panicle

Papilionaceous A type of flower with a banner, 2 wing petals, and 2 partially fused or fused keel petals

Pedicel The stalk of a single flower

Pedicellate Borne on a pedicel

Peduncle The stalk of a flower cluster or of one flower when it is the only one in the inflorescence

Pendulous Suspended or hanging downward; drooping

Perennial Lasting more than 2 years; applied to plants living 2 or more years

Perfect Applied to flowers with both functional stamens and pistils

Perianth Collective term referring to both the corolla and calyx

Persistent Not falling away

Petal A part or member of the corolla, often brightly colored

Petaliferous Bearing petals

Petiolate With a petiole

Petiole The stalk of a leaf

Petiolule The stalk of a leaflet in a compound leaf

Phyllode Leaflike petiole serving as the blade

Pilose With long, soft, straight hairs

Pilosulous Minutely pilose

Pinnae Primary divisions of a pinnate leaf

Pinnate Describing a compound leaf with 2 rows of leaflets on opposite sides of the axis

Pistil The female part of the flower consisting of the stigma, style, and ovary

Pistillate Applied to flowers bearing pistils only

Plumose Feathery; with fine, long pubescence

Pod A dry dehiscent fruit

Poly- A prefix meaning many

Polygamous Having bisexual and unisexual flowers on the same plant

Polygamodioecious Dioecious plants having some perfect flowers

Prickle A small spinelike structure produced from the epidermis or bark

Prostrate Lying flat on the surface of the ground

Proximal Nearest point of attachment

Puberulent Diminutive of pubescence; hairs hardly visible

Pubescence The hairs on the epidermis of a plant

Pubescent Covered with hair

Pulvinate Cushion-shaped

Punctate Having dots; usually small glandular pits

Pustulate Having minute blisters or blisterlike projections

Raceme A simple inflorescence in which the flowers are pediceled on a rachis

Rachis The main axis of a spike, raceme, or other inflorescence

Recurved Curved outwards or backwards; curved downwards

Reflexed Abruptly recurved

Regular Describing flowers that are radially symmetrical; having all the parts in each series alike in size and shape

Reniform Kidney- or bean-shaped

Resinous Producing or containing viscous substances such as resin or amber

Reticulate In the form of a network; netted

Retrorse Pointing backward or downward toward the base

Retuse Somewhat notched at the blunt or rounded apex

Rhizobial Containing nitrogen-fixing bacteria

Rhizomatous Having rhizomes

Rhizome An underground stem, usually lateral and rooting at the nodes

Rhombic, rhomboid Shaped like a diamond; shaped like an equilateral parallelogram

Rib A prominent vein or nerve

Rosette A whorl of leaves, usually basal

Rotund Nearly circular

Rudimentary Imperfectly developed

Rugose With a wrinkled surface

Rugulose Somewhat wrinkled

Saccate Sac- or bag-shaped

Sagittate Triangular or arrow-shaped with the lobes turned downward

Scabrous Rough to the touch

Scale Any thin and dry, appressed organ; usually a modified leaf

Scape A leafless peduncle arising from the ground or from a basal whorl of leaves and bearing 1 or more flowers

Scapose Bearing flowers on a scape

Scarification Process that cuts or breaks the seed coat to allow water and gases to enter the seeds, used to enhance germination

Scarious Thin, dry, membranous, and somewhat translucent

Secund Having flowers or leaves directed to one side

Semi- A prefix meaning half

Sepal One division of the calyx

Septate Divided by partitions

Sericeous Silky with soft, appressed hairs

Serrate Saw-toothed margins with teeth pointing toward the apex

Serrulate Minutely serrate

Sessile Without a stalk or pedicel

Setaceous Having bristlelike hairs

Simple Not compound; not branched

Sinuous With a strongly wavy margin

Sinus Indentation between 2 lobes or segments

Spatulate Shaped like a spatula or spoon; broader above than below

Spicate Spikelike

Spike An unbranched inflorescence in which the flowers are sessile on the rachis

Spine A small, rigid, and sharp-pointed outgrowth

Spur A slender, hollow projection from a flower

Stamen The male part of the flower consisting of the anther and filament

Staminate Describing flowers containing only stamens

Standard The upper petal of a papilionaceous corolla; banner

Stigma The part of the pistil that is receptive to pollen

Stipe The stalk of a pistil; in general a stalk that supports an organ

Stipel The stipule of a leaflet

Stipellate Having stipels

Stipitate Borne on a stipe

Stipule Appendages, usually leaf-like, occurring in pairs, 1 on either side of the petiole base

Stolon A horizontal stem that roots at the nodes; runner

Stratification A cold, moist treatment of seeds used to enhance germination

Striate Marked with slender longitudinal grooves or lines; appearing striped

Strigose Rough with short, stiff hairs or bristles, often swollen at the base

Strigulose Minutely strigose

Style That part of the pistil which bears the stigma

Sub- Prefix meaning somewhat, slightly, or in less degree

Subtend To be situated closely beneath

Subterranean Located underground

Subulate Awl-shaped; sharp-pointed from a broadened base

Suffrutescent Having a shrubby base

Suffruticose Plants with woody bases and herbaceous above

Suffuse To spread through or over with color

Sulcate Deeply grooved or furrowed lengthwise

Sulcus A groove or furrow

Suppressed Failing to develop

Suture A line of dehiscence in a dry fruit

Symmetrical All parts regular in number, size, and shape

Taproot The main vertical root, often stout

Tendril A slender, sometimes branching, coiled structure used by some plants for clinging

Terete Cylindrical; circular in cross-section

Terminal Borne at, or belonging to, the summit or extremity

Tomentose Densely pubescent with matted and tangled hairs

Transverse Perpendicular to the longitudinal axis of a structure

Tri- A prefix meaning 3

Triangular 3-sided; 3-angled

Triquetrous 3-angled with concave faces between

Trimorphic Having, or occurring in, 3 different forms

Truncate Ending abruptly; appearing to be cut off at the end

Tuber A thickened portion of an underground stem with numerous buds

Turbinate Top-shaped; inversely conical

Turgid Rigid or swollen with water

Umbel A simple inflorescence with pedicels radiating from a common point, flat-topped or rounded

Uncinate Hooked at the tip; in the form of a hook

Uncinulate Minutely uncinate

Uni- Prefix meaning 1

Valvate Opening by valves; meeting by margins but not overlapping

Valve 1 of the ovary wall segments into which a longitudinally dehiscent fruit splits

Vein A single branch of the vascular system of a plant

Venation The arrangement of veins

Ventral The upper or inner surface of an organ

Verrucose Warty; covered with small projections

Villous With long, soft hairs that are not matted

Villosulous Minutely villous

Virgate Long, straight, and slender; wand-shaped

Whorl A cluster of leaves or branches around the axis

Wing The lateral petal of a papilionaceous flower; a thin, membranaceous extension of an organ

Woolly Covered with long, entangled soft hairs

Zygomorphic Capable of division into symmetrical halves only in 1 plane; irregular

AUTHORITIES

Ait.	William Aiton (1731–1793), English botanist and Royal Gardener at Kew.
Arcang.	Giovanni Arcangeli (1840–1921), Italian botanist.
Barneby	Rupert Charles Barneby (1911–), botanist with the New York Botanical Garden.
Bart.	William Paul Crillon Barton (1786–1856), professor of botany and United States Navy surgeon.
Bartal.	Biagio Bartalini (1746–1822), Italian professor of botany.
Bates	Rev. John Mallery Bates (1846–1930), Nebraska clergyman and botanist.
Benth.	George Bentham (1800–1884), English taxonomist and president of the Linnaean Society.
Bisch.	Gottlieb Wilhelm Bischoff (1797–1854), German professor of botany.
Blake	Sidney Fay Blake (1892–1959), scientist and author with the United States Department of Agriculture.
Blank.	Joseph William Blankinship (1862–1938), botanist with Montana State University.
Boynt.	Frank Ellis Boynton (1859-?), botanical collector for the herbarium of the Biltmore Estates, Biltmore, North Carolina.
Br., R.	Robert Brown (1773–1858), botanist, British Museum.
Brew.	William Henry Brewer (1828–1910), botanist and geologist, California State Geological Survery.
Britt.	Nathaniel Lord Britton (1859–1934), director-in-chief of the New York Botanical Garden.
Britt. & Rose	Nathaniel Lord Britton (1859–1934), director-in-chief of the New York Botanical Garden; and Joseph Nelson Rose (1862–1928), botanist, United States National Herbarium.
B.S.P.	Nathaniel Lord Britton (1859–1934), director-in-chief of the New York Botanical Garden; Emerson Ellick Sterns (1846–1926); Justus Ferdinand Poggenberg (1840–1893).
Buckl.	Samuel Botsford Buckley (1809–1884), naturalist and state geologist of Texas.
Bullock	Arthur A. Bullock (1906–), English botanist at Kew.
Butt. & St. John	Fredrick King Butters (1878–1945), professor of botany, University of Minnesota; and Harold St. John (1892–), professor of botany, Washington State University and University of Hawaii.
Cav.	Antonio Jose Cavanilles (1745–1804), Spanish professor of botany and director of the botanic gardens in Madrid.
Clem. & Clem.	Frederic Edward Clements (1874–1945), plant ecologist with the University of Nebraska and the Carnegie Institute; Edith Gertrude (Schwartz) Clements (1877–1971), wife of Frederic, ecologist, and author.
Cory	Victor Louis Cory (1880–1964), botanist, Southern Methodist University.
Darl.	Josephine Darlington (1905–), botanist, Missouri Botanical Garden.

DC.	Augustin Pyramus de Candolle (1778–1841), Swiss botanist and professor of botany.
DC., A.	Alphonse Louis Pierre Pyramus de Candolle (1806–1893), son of Augustin, author of botanical books.
Desv.	Nicaise Auguste Desvaux (1784–1856), French botanist and professor of botany.
Dietr.	Friedrich Gottlieb Dietrich (1768–1850), garden director at Eisenach, Germany.
Don, G.	George Don (1786–1856) English botanist and plant collector
Dougl.	David Douglas (1798–1834), Scottish botanical collector in northwestern United States of America for the Royal Horticultural Society.
Dumont	George Louis Marie Dumont de Courset (1746–1824), French agronomist.
Eat.	Amos Eaton (1776–1842), American botanist, produced the first botanical manual in America with descriptions in English.
Eifert	Imre Janos Eifert (1934–), formerly with the Department of Botany, University of Texas.
Ell.	Stephen Elliot (1771–1830), American botanist and professor in South Carolina.
Engelm.	George Engelmann (1809–1884), physician and botanist in St. Louis, Missouri.
Fabr.	Philipp Conrad Fabricius (1714–1774), German physician and botanist.
Fern.	Merritt Lyndon Fernald (1873–1950), plant taxonomist, plant geographer, director of Gray Herbarium (1937–1947).
Fisch.	Friedrich Ernst Ludwig von Fischer (1782–1854), director of St. Petersburg (Russia) Botanical Garden.
Gray, A.	Asa Gray (1810–1888), professor of botany at Harvard University.
Greene	Edward Lee Greene (1842–1915), professor of botany at the University of California, Smithsonian Institution, and Catholic University of America.
Greenm. & Larisey	Jesse More Greenman (1867–1951), curator of the herbarium, Missouri Botanical Garden (1913–1948); and Mary Maxine Larisey (1909–), School of Pharmacy, Medical College of South Carolina.
H. & A.	Sir William Jackson Hooker (1785–1865), English botanist and director of the Royal Botanic Gardens at Kew (1841–1865); and George Arnold Walker Arnott (1799–1869), Scottish botanist.
Heller	Amos Arthur Heller (1867–1944), Pennsylvania botanist and collector of western plants.
Herm.	Frederick Joseph Hermann (1906–), scientist with United States Department of Agriculture and the Forest Service.
Hitchc., C. L.	Charles Leo Hitchcock (1902–), botanist with the University of Washington.
Hook.	Sir William Jackson Hooker (1785–1865), English botanist and director of the Royal Botanic Gardens at Kew (1841–1865).
Hornem.	Jens Wilken Hornemann (1770–1841), professor of botany, Denmark.
Isley	Duane Isley (1918–), professor of botany at Iowa State University.
Jones, M. E.	Marcus Eugene Jones (1852–1934), collector of Great Basin plants and mining consultant.
Kearn.	Thomas Henry Kearney (1874–1956), taxonomist with the United States Department of Agriculture.

Kell.	Albert Kellogg (1813–1887), physician, botanist, and founder of the California Academy of Sciences.
Kelso	Leon Hugh Kelso (1907–), scientist with the United States Biological Survey in Wyoming and Colorado.
Ker	John Bellenden Ker (1764–1842), British botanist, also known as John Ker Bellenden and John Gawler.
Koch	Wilhelm Daniel Joseph Koch (1771–1849), German professor of botany and author.
Koch, K.	Karl Heinrich Emil Koch (1809–1879), German traveler, dendrologist, and professor.
Ktze. O.	Carl Ernst Otto Kuntze (1843–1907), German botanist.
L.	Carolus Linnaeus (1707–1778), botanist at Uppsala, Sweden; author of *Species Plantarum* (1753) upon which botanical nomenclature is based, the "Father of Taxonomy".
Lam.	Jean Baptiste Antoine Pierre Monet, Chevalier de Lamarck (1744–1829), French naturalist and author.
Larisey	Mary Maxine Larisey (1909–), School of Pharmacy, Medical College of South Carolina.
Lehm.	Johann Georg Christian Lehmann (1792–1860), German botanist and director of the botanic garden in Hamburg.
Lindl.	John Lindley (1799–1865), English professor of botany, horticulturist, and author.
Loud.	John Claudius Loudon (1783–1843), English horticulturist and author.
Lucanus	Anonymous citizen of Lucca (1753).
Mack. & Bush	Kenneth Kent Mackenzie (1877–1934), New York City attorney and botanist; Benjamin Franklin Bush (1858–1937), Missouri storekeeper and amateur botanist.
MacM.	Conway MacMillan (1867–1929), Minnesota state botanist.
Maxim.	Carl Johann Maximowicz (1827–1891), Russian botanist and director of the botanic garden in St. Petersburg, also known as Karl Ivanovich Maksimovich.
Medic.	Friedrich Kasimir Medicus (1736–1808), German botanist.
Michx.	Andre Michaux (1746–1802), French botanist and explorer of North America.
Mill., P.	Philip Miller (1691–1771), British gardner and author.
Moench	Conrad Moench (1744–1805), German professor of botany.
Muhl.	Gotthilf Heinrich Ernst Muehlenberg (1753–1815), German-educated Lutheran minister and pioneer botanist in Pennsylvania, also known as Henry Muhlenberg and Heinrich Ludwig Muhlenberg.
Nels., A.	Aven Nelson (1859–1952), botanist and president of the University of Wyoming.
Nieuw. & Lunnell	Julius Aloysius Arthur Nieuwland (1878–1936), professor of botany and organic chemistry, Notre Dame University; Joel Lunnell (1851–1920), North Dakota botanist.
Nutt.	Thomas Nuttall (1786–1859), English-American naturalist and botanist who collected in western America.
Ort.	Casimiro Gomez Oretga (1740–1818), Spanish botanist and director of the botanical garden in Madrid.
Pall.	Peter Simon Pallas (1741–1811), German botanist and author.

Palm., E. J.   Ernest Jesse Palmer (1875–1962), collector for the Arnold Arboretum and Missouri Botanical Garden.

Peck   Morton Eaton Peck (1871–1959), author and botanist with Williamette University.

Pers.   Christiaan Hendrik Persoon (1761–1836), South African-French botanist and author.

Phillips   Lyle L. Phillips (1923–   ), botanist with the University of Washington.

Piper   Charles Vancouver Piper (1867–1926), professor of botany and zoology with Washington State University.

Pollard & Ball   Charles Louis Pollard (1872–1945), plant collector and librarian from Vermont; Carlton Roy Ball (1873–1958), agronomist with the United States Department of Agriculture.

Pursh   Frederick Traugott Pursh (1774–1820), author and botanist, born in Saxony and lived in Philadelphia, collected extensively in North America.

Raf.   Constantine Samuel Rafinesque (1783–1840), naturalist, born in Constantinople and lived in Kentucky.

Richards.   Sir John Richardson (1787–1865), Scottish biologist attached to an expedition to arctic North America.

Roth   Albrecht Wilhelm Roth (1757–1834), German botanist and physician.

Rydb.   Per Axel Rydberg (1860–1931), Swedish-born, American botanist, author, and curator of the New York Botanical Garden.

Scheele   Georg Heinrich Adolf Scheele (1808–1864), German botanist who worked with Texas plants.

Schneid.   Camillo Karl (formerly Carl Camillo) Schneider (1876–1951), German and Austrian dendrologist.

Schreb.   Johann Christian Daniel von Schreber (1739–1810), German botanist and professor.

Schub.   Bernice Giduz Schubert (1913–   ), American botanist with Harvard University and the United States Department of Agriculture.

Schulz   Otto Eugen Schulz (1874–1936), German taxonomist.

Schum., K.   Karl Moritz Schumann (1851–1904), German botanist and curator of the herbarium in Berlin.

Scop.   Johann Anton Scopoli (1723–1788), Austrian botanist and professor, also known as Giovanni Antonio Scopoli.

Shafer   John Adolf Shafer (1863–1918), botanist with New York Botanical Museum.

Sheld.   Edmund Perry Sheldon (1869-?) botanical author, resident of Minnesota and Oregon.

Shinners   Lloyd Herbert Shinners (1918–1971), Canadian-born botanist, professor of botany at Southern Methodist University.

Shuttlew.   Robert James Shuttleworth (1810–1874), English botanist.

Sibth.   John Sibthorp (1758–1796), professor of botany, Oxford, England.

Sm. & Rydb.   Jared Gage Smith (1866–1925), agrostologist with the United States Department of Agriculture; Per Axil Rydberg (1860–1931), Swedish-born, American botanist and curator of the New York Botanical Garden.

Small, J.   John Kunkel Small (1869–1938), American botanist, author, and curator of the New York Botanical Garden.

Spreng.   Curt Polykarp Joachim Sprengel (1766–1833), German professor of medicine and botany at Halle.

Standl.	Paul Carpenter Standley (1884–1963), curator United States National Herbarium and Chicago Natural History Museum.
Sw.	Olof Peter Swartz (1760–1818), Swedish professor of botany and author.
Sweet	Robert Sweet (1783–1835), English horticulturist and author.
Swezey	Goodwin Deloss Swezey (1851–1934), professor of astronomy and Nebraska naturalist.
T. & G.	John Torrey (1796–1873), American physician and professor of botany and chemistry; Asa Gray (1810–1888), professor of botany at Harvard University.
Ten.	Michele Tenore (1780–1861), Italian professor of botany.
Thieret	John William Thieret (1926–), professor at Northern Kentucky University.
Thunb.	Carl Peter Thunberg (1743–1828), Swedish professor of botany and author.
Tidest.	Ivar T. Tidestrom (1864–1956), Swedish-born professor of botany at the Catholic University of America.
Torr.	John Torrey (1796–1873), American physician and professor of botany and chemistry.
Torr. & Frem.	John Torrey (1796–1873), American physician and professor of botany and chemistry; John Charles Fremont (1813–1890), soldier, explorer, and presidential candidate.
Trel.	William Trelease (1857–1945), director of the Missouri Botanical Garden, professor of botany at the University of Illinois, and first president of the Botanical Society of America.
Turner, B. L.	Billie Lee Turner (1925–), professor of botany at the University of Texas.
Vent.	Etienne Pierre Ventenat (1757–1808), French professor of botany.
Waldst. & Kit.	Count Franz de Paula Adam von Waldstein-Wartemberg (1759–1823), botanical author; Paul Kitaibel (1757–1817), professor of botany and chemistry, Budapest.
Wats., S.	Sereno Watson (1826–1892), botanist and assistant to Asa Gray.
Wemple	Don Kimberly Wimple (1929-), at Iowa State University in the mid-1960's.
Willd.	Carl Ludwig von Willldenow (1765–1812), German botanist and director of the Berlin Botanical Garden.
Wood	Alphonso Wood (1810–1881), principal and botanical author, New York.
Woot.	Elmer Otis Wooten (1865–1945), professor of botany at New Mexico State University, later with the United States Department of Agriculture.
Woot. & Standl.	Elmer Otis Wooten (1865–1945), professor of botany at New Mexico State University, later with the United States Department of Agriculture; Paul Carpenter Standley (1884–1963), curator United States National Herbarium and Chicago Natural History Museum.

SELECTED REFERENCES

Agricultural Research Service. 1970. Selected weeds of the United States. United States Department of Agriculture, Washington, D.C.

Allen, O. N., and E. K. Allen. 1981. The Leguminosae. University of Wisconsin Press, Madison.

Alley, H. P., and G. A. Lee. 1969. Weeds of Wyoming. Bulletin 498. Agricultural Experiment Station, University of Wyoming, Laramie.

Apagar, A. C. 1892. Trees of the northern United States. American Book Company, New York.

Apagar, A. C. 1910. Ornamental shrubs of the United States. American Book Company, New York.

Atchison, E. 1949. Studies in the Leguminosae. IV. Chromosome numbers and geographical relationships of miscellaneous Leguminosae. Journal of the Elisha Mitchell Science Society 65:118–122.

Bailey Hortorium. 1976. Hortus third. Macmillan Publishing Company, New York.

Bare, J. E. 1979. Wildflowers and weeds of Kansas. Regents Press of Kansas, Lawrence.

Barkley, T. M. 1983. Field guide to the common weeds of Kansas. University Press of Kansas, Lawrence.

Barneby, R. C. 1952. A revision of the North American species of *Oxytropis* DC. Proceedings of the California Academy of Sciences 27:117–312.

Blake, S. F. 1924. Notes on American lespedezas. Rhodora 26:25–34.

Budd, A. C. 1957. Wild plants of the Canadian prairies. Publication 983. Canada Department of Agriculture, Saskatchewan.

Campbell, J. B., R. W. Lodge, and A. C. Budd. 1956. Poisonous plants of the Canadian prairies. Publication 900. Canada Department of Agriculture, Ottawa.

Cheeke, P. R., and L. R. Shull. 1985. Natural toxicants in feeds and poisonous plants. AVI Publishing Company, Westport, Connecticut.

Clark, L. 1973. Wild flowers of British Columbia. Gray's Publishing Limited, Sidney, British Columbia.

Coon, N. 1979. Using plants for healing. Rodale Press, Emmaus, Pennsylvania.

Correll, D. S., and M. C. Johnston. 1970. Manual of the vascular plants of Texas. Texas Research Foundation, Renner.

Currath, R., A. Smreciu, and M. Van Dyk. 1983. Prairie wildflowers. University of Alberta, Edmonton.

Curtin, L. S. M. 1947. Healing herbs of the Upper Rio Grande. University of New Mexico, Santa Fe.

Davison, V. E. 1958. A summary and reclassification of bobwhite foods. Journal of Wildlife Management 22:437–439.

Dayton, W. A. 1931. Important western browse plants. Miscellaneous Publication 101. United States Department of Agriculture, Washington, D.C.

Densmore, F. 1928. Uses of plants by the Chippewa Indians. Annual Report of the Bureau of American Ethnology to the Secretary of the Smithsonian Institution, 1926–1927 44:275–397.

Dirr, M. A. 1983. Manual of woody landscape plants. Stipes Publishing Company, Champaign, Illinois.

Dunn, D. B., and J. M. Gillett. 1966. The lupines of Canada and Alaska. Monograph No. 2. Canada Department of Agriculture, Ottawa.

Elias, T. S. 1980. Trees of North America. Van Nostrand Reinhold Company, New York.

Evers, R. A., and R. P. Link. 1972. Poisonous plants of the Midwest & their effects on livestock. Special Publication 24. University of Illinois, Urbana-Champaign.

Farrell, E. K., P. E. Collins, and W. G. Marksom. 1965. Trees of South Dakota. Extension Circular 566. Cooperative Extension Service, South Dakota State University, Brookings.

Fassett, N. C. 1939. The leguminous plants of Wisconsin. University of Wisconsin Press, Madison.

Fernald, M. L. 1950. Gray's manual of botany. The Macmillan Company, New York.

Forest Service. 1937. Range plant handbook. Forest Service. United States Department of Agriculture, Washington, D.C.

Frahm-Leliveld, J. A. 1953. Some chromosome numbers in tropical leguminous plants. Euphytica 2:46–48.

Gambill, W. G., Jr. 1953. The Leguminosae of Illinois. Illinois Biological Monographs. Volume XXII. University of Illinois Press, Urbana.

Gates, F. C. 1930. Principal poisonous plants in Kansas. Technical Bulletin 25. Agricultural Experiment Station, Kansas State Agricultural College, Manhattan.

Gates, F. C. 1941. Weeds in Kansas. Kansas State Board of Agriculture, Topeka.

George, E. J. 1953. Tree and shrub species for the Northern Great Plains. Circular 912. United States Department of Agriculture, Washington, DC.

Gilmore, M. R. 1919. Uses of plants by the Indians of the Missouri River region. Annual Report of the Bureau of American Ethnology 1911–12:43–154.

Gilmour, J. S. L. 1980. International code of nomenclature of cultivated plants. International Association for Plant Taxonomy, Utrecht, Netherlands.

Gleason, H. A. 1952. New Britton and Brown illustrated flora of the northeastern United States and adjacent Canada. Lancaster Press, Lancaster, Pennsylvania.

Gould, F. W. 1975. Texas plants—a checklist and ecological summary. Miscellaneous Publication 585. Texas Agricultural Experiment Station, College Station.

Graham, E. H. 1941. Legumes for erosion control and wildlife. Miscellaneous Publication No. 412. United States Department of Agriculture, Washington, D.C.

Grelen, H. E., and R. H. Hughes. 1984. Common herbaceous plants of southern forest range. Research Paper SO-210. Southern Forest Experiment Station, United States Department of Agriculture, New Orleans.

Grinnell, G. B. 1962. The Cheyenne Indians—their history and ways of life. Volume 2. Cooper Square Publishing Company, New York.

Hafenrichter, A. L., L. A. Mullen, and R. L. Brown. 1949. Grasses and legumes for soil conservation in the Pacific Northwest. Miscellaneous Publication 678. United States Department of Agriculture, Washington, D.C.

Hafenrichter, A. L., J. L. Schwendiman, H. L. Harris, R. S. McLauchlan, and H. W. Miller. 1968. Grasses and legumes for soil conservation in the Pacific Northwest and Great Basin states. Handbook 339. United Stated Department of Agriculture, Washington, DC.

Harrington, H. D. 1954. Manual of the plants of Colorado. Sage Books, Denver.

Hart, J. and J. Moore. 1976. Montana-native plants and native people. Montana Historical Society, Helena.

Herman, F. J. 1953. A botanical synopsis of the cultivated clovers (*Trifolium*). Monograph No. 22. United Stated Department of Agriculture, Washington, DC.

Herman, F. J. 1966. Notes on western range forbs: Cruciferae through Compositae. Handbook 293. United States Department of Agriculture, Washington, DC.

Hitchcock, C. L. 1952. A revision of the North American species of *Lathyrus*. University of Washington Biology 15:1–104.

Hitchcock, C. L., and A. Cronquist. 1973. Flora of the Pacific Northwest. University of Washington Press, Seattle.

Irvin, H. S. 1961. Roadside flowers of Texas. University of Texas Press, Austin.

Isley, D. 1953. *Desmodium paniculatum* (L.) DC. and *D. viridiflorum* (L.) DC. American Midland Naturalist 49:920–933.

Isley, D. 1983. *Astragalus* (Leguminosae: Papilionoideae) I.: Keys to United States species. Iowa State Journal of Research 58:1–172.

Isley, D. 1984. *Astragalus* L. (Leguminosae: Papilionoideae) II: Species summary A-E. Iowa State Journal of Research 59:97–216.

Isley, D. 1985. Leguminosae of United States, *Astragalus* L.: III. Species summary F-M. Iowa State Journal of Research 60:179–322.

Isley, D. 1986. Leguminosae of the United States. *Astragalus* L.: IV. Species summary N-Z. Iowa State Journal of Research 61:153–296.

Jaeger, E. C. 1950. A source-book of biological names and terms. Charles C. Thomas, Springfield, Illinois.

James, L. F. 1981. Syndromes of *Astragalus* poisoning in livestock. Journal of the American Veterinary Medical Association 158:614–618.

James, L. F., R. F. Keeler, A. E. Johnson, M. C. Williams, E. H. Cronin, and J. D. Olsen. 1980. Plants poisonous to livestock in the western states. Agriculture Information Bulletin 415. United States Department of Agriculture, Washington, D.C.

Johnson, J. R., and J. T. Nichols. 1982. Plants of South Dakota grasslands. Bulletin 566. Agricultural Experiment Station, South Dakota State University, Brookings.

Jones, G. N. 1971. Flora of Illinois. University of Notre Dame, Notre Dame, Indiana.

Kartesz, J. T., and R. Kartesz. 1980. A synonymized checklist of vascular flora of the United States, Canada, and Greenland. Volume II. The Biota of North America. University of North Carolina Press, Chapel Hill.

Keeler, R. F., K. R. Van Kampen, and L. F. James (Eds.). 1978. Effects of poisonous plants on livestock. Academic Press, New York.

Kelsey, H. P., and W. A. Dayton. 1942. Standardized plant names. Horace McFarland Company, Harrisburg, Pennslyvania.

Kinch, R. C., L. Wrage, and R. A. Moore. 1975. South Dakota weeds. Cooperative Extension Service, South Dakota State University, Brookings.

Kingsbury, J. M. 1964. Poisonous plants of the United States and Canada. Prentice-Hall, Englewood Cliffs, New Jersey.

Kirk, D. R. 1975. Wild edible plants of the western United States. Naturegraph, Healdsburg, California.

Kreig, M. 1964. Green medicine. Rand McNally, Chicago.

Krochmal, A., and C. Krochmal. 1973. A guide to the medicinal plants of the United States. Quadrangle/New York Times Book Company, New York.

Larisey, M. M. 1940. A monograph of the genus *Baptisia*. Annals of the Missouri Botanical Garden 27:119–244.

Leininger, W. C., J. E. Taylor, and C. L. Wambolt. 1977. Poisonous range plants in Montana. Bulletin 348. Cooperative Extension Service, Montana State University, Bozeman.

Lewis, W. H., and P. F. Elvin-Lewis. 1977. Medical botany. John Wiley & Sons, New York.

Lommasson, R. C. 1973. Nebraska wild flowers. University of Nebraska Press, Lincoln.

Looman, J., and K. F. Best. 1979. Budd's flora of the Canadian Prairie Provinces. Publication 1662. Canada Agriculture, Ottowa.

Martin, A. C., H. S. Zim, and A. L. Nelson. 1951. American wildlife and plants. McGraw-Hill Book Company, New York.

McGregor, R. L., and T. M. Barkley (Eds.). 1977. Atlas of the flora of the Great Plains. Iowa State University Press, Ames.

McGregor, R. L., and T. M. Barkley (Eds.). 1986. Flora of the Great Plains. University Press of Kansas, Lawrence.

Mohlenbrock, R. H. 1957. A revision of the genus *Stylosanthes*. Annals of the Missouri Botanical Garden 44:299–355.

Morre, J. A. 1936. The vascular anatomy of the flower in the papilionaceous Leguminosae. I. American Journal of Botany 23:279–290.

Nebraska Statewide Arboretum. 1982. Common and scientific names of Nebraska plants. Publication 101. Nebraska Statewide Arboretum, Lincoln.

Nelson, R. A. 1968. Wild flowers of Wyoming. Bulletin 490. Cooperative Extension Service, University of Wyoming, Laramie.

Over, W. H. 1932. Flora of South Dakota. University of South Dakota, Vermillion.

Owensby, C. E. 1980. Kansas prairie wildflowers. Iowa State University Press, Ames.

Phillips Petroleum Company. 1963. Pasture and range plants. Bartlesville, Oklahoma.

Pool, R. J. 1971. Handbook of Nebraska trees. Nebraska Conservation Bulletin 32. Conservation and Survey Division, University of Nebraska, Lincoln.

Porter, C. L. 1960. Wyoming trees. Circular 164R. Cooperative Extension Service, University of Wyoming, Laramie.

Preston, R. J., Jr. 1976. North American trees. Iowa State University Press, Ames.

Radford, A. E., W. C. Dickison, J. R. Massey, and C. R. Bell. 1974. Vascular plant systematics. Harper & Row, New York.

Rehder, A. 1977. Manual of cultivated trees and shrubs. Macmillan Publishing Company, New York.

Rogers, D. J. 1980. Edible, medicinal, useful, and poisonous wild plants of the Northern Great Plains—South Dakota region. Buechel Memorial Lakota Museum, St. Francis.

Rogers, D. J. 1980. Lakota names and traditional uses of native plants by Sicangu (Brule) people in the Rosebud area, South Dakota. Rosebud Educational Society, St. Francis.

Rogler, G. A. 1967. Grass and legume introductions in the northern Great Plains. Proceedings of the Great Plains Agricultural Council, Texas A&M University, College Station.

Rosendahl, C. O. 1955. Trees and shrubs of the upper Midwest. University of Minnesota Press, St. Paul.

Rydberg, P. A. 1906. Flora of Colorado. Bulletin 100. Agricultural Experiment Station, Colorado Agricultural College, Fort Collins.

Rydberg, P. A. 1932. Flora of the prairies and plains of central North America. New York Botanical Gardens, New York.

Salac, S. S., P. N. Jensen, J. A. Dickerson, and R. W. Gray, Jr. 1978. Wildflowers for Nebraska landscapes. Miscellaneous Publication 35. Nebraska Agricultural Experiment Station, Lincoln.

Saunders, C. F. 1934. Useful wild plants of the United States and Canada. Robert M. McBride & Company, New York.

Schopmeyer, C. S. (Technical Coordinator). 1974. Seeds of woody plants in the United States. Handbook 450. United States Department of Agriculture, Washington, DC.

Smith, H. H. 1928. Ethnobotany of the Meskwaki Indians. Public Museum, City of Milwaukee Bulletin 4:175–326.

Sperry, O. E., J. W. Dollahite, G. O. Hoffman, and B. J. Camp. 1977. Texas plants poisonous to livestock. Cooperative Extension Service, Texas A&M University, College Station.

Standley, P. C. 1922. Trees and shrubs of Mexico (Leguminosae). Contributions of the United States National Herbarium 23:348–515.

State of Nebraska. 1962. Nebraska weeds. Bulletin 101-R. Weed and Seed Division, Department of Agriculture and Inspection, Lincoln.

Stephens, H. A. 1973. Woody plants of the north central states. University Press of Kansas, Lawrence.

Stevens, O. A. 1950. Handbook of North Dakota plants. North Dakota Agricultural College, Fargo.

Steyermark, J. A. 1963. Flora of Missouri. Iowa State University Press, Ames.

Stroh, J. R., A. E. Carlson, and W. J. Seamands. 1972. Management of Lutana cicer milkvetch for hay, pasture, seed, and conservation uses. Research Journal 66. Soil Conservation Service, Montana State University, and University of Wyoming, Laramie.

Stubbendieck, J., S. L. Hatch, and K. J. Hirsch. 1986. North American range plants. University of Nebraska Press, Lincoln.

Tatschl, A. K. 1970. A taxonomic and life history study of *Psoralea tenuiflora* and *Psoralea floribunda* in Kansas. Ph.D. Dissertation, University of Kansas, Lawrence.

Texas Forest Service. 1963. Forest trees of Texas. Texas Forest Service, Texas A&M University, College Station.

Thomas, J. O., and L. J. Davies. 1964. Common British grasses and legumes. Longmans, Green and Company, London.

Thornburg, A. A. 1982. Plant materials for use on surface-mined lands in arid and semi-arid regions. Technical Publication 157. Soil Conservation Service, United States Department of Agriculture, Washington, DC.

Turner, B. L. 1959. The legumes of Texas. University of Texas Press, Austin.

Van Bruggen, T. 1971. Wildflowers of the northern plains and Black Hills. Bulletin 3. Badlands Natural History Association, Wall, South Dakota.

Van Bruggen, T. 1985. The vascular plants of South Dakota. Iowa State University Press, Ames.

Vance, F. R., J. R. Jowsey, and J. S. McLean. 1984. Wildflowers of the northern Great Plains. University of Minnesota Press, Minneapolis.

Viertel, A. T. 1970. Trees, shrubs, and vines. Syracuse University Press, Syracuse.

Vines, R. A. 1960. Trees, shrubs, and woody vines of the Southwest. University of Texas Press, Austin.

Vogel, V. J. 1970. American Indian medicine. University of Oklahoma Press, Norman.

Waterfall, U. T. 1962. Keys to the flora of Oklahoma. Oklahoma State University Press, Stillwater.

Weiner, M. A. 1980. Earth medicine—earth food. Macmillan Publishing Company, New York.

Whitson, T. D. (Ed.). 1987. Weeds and poisonous plants of Wyoming and Utah. Cooperative Extension Service, University of Wyoming, Laramie.

Wilber, R. L. 1963. The leguminous plants of North Carolina. Technical Bulletin No. 151. North Carolina Agricultural Experiment Station, Durham.

Williams, K. 1977. Eating wild plants. Mountain Press Publishing Company, Missoula, Montana.

INDEX

LIST OF FIGURES

327

34. *Astragalus lotiflorus* 0.7X, flower 2.8X, leaflets 2.6X, seed 4.8X, legume 1.7X, inflorescence in fruit 0.7X.

35. *Astragalus missouriensis* 0.7X, flower 3.2X, inflorescence in fruit 0.7X, legume 1.9X, dehisced legume 1.5X, seed 5.3X, leaflet 2.3X.

36. *Astragalus mollissimus* 0.7X, flower 2.8X, leaflets 2.1X, seed 7.7X, legume 2.6X, inflorescence in fruit 0.7X.

37. *Astragalus nuttallianus* 0.7X, flower 6.3X, seed 4.7X, legume 1.8X, leaflet 2.8X.

38. *Astragalus pectinatus* 0.7X, leaflet 2.0X, flower 3.2X, legume 2.1X, seed 4.7X.

39. *Astragalus plattensis* 0.7X, flower 3.9X, seed 2.2X, legume 1.8X, leaflets 4.9X.

40. *Astragalus racemosus* 0.7X, flower 3.2X, leaflets 1.7X, inflorescence in fruit 0.7X, legume 2.1X, seed 4.1X.

41. *Astragalus sericoleucus* 0.7X, flower 5.7X, legume 5.0X, seed 4.6X, leaf 3.4X.

42. *Astragalus spatulatus* 0.7X, flower 5.2X, leaf 3.4X, seed 10.5X, legume 4.3X, inflorescence in fruit 0.7X.

43. *Astragalus tenellus* 0.7X, flower 4.6X, leaflets 4.3X, legume 3.8X, stipules 8.4X, seed 5.9X.

44. *Baptisia australis* 0.7X, legume 1.3X, seed 3.0X, leaflet 1.9X.

45. *Baptisia bracteata* 0.7X, flower 1.5X, legume 1.4X, seed 4.4X.

46. *Baptisia lactea* 0.7X, flower 2.0X, legume 1.0X, seed 3.1X.

47. *Baptisia tinctoria* 0.7X, flower 3.2X, legume 2.2X, seed 5.1X, stipule 2.8X.

48. *Caragana arborescens* 0.7X, flower 2.4X, seed 2.0X, leaflet 1.7X.

49. *Coronilla varia* 0.7X, flower 3.5X, leaflet 2.0X, seed 3.7X, loment 1.5X, inflorescence in fruit 0.9X.

50. *Crotalaria sagittalis* 0.7X, flower 7.4X, leaf 1.3X, seed 4.9X, legume 2.0X.

51. *Dalea aurea* 0.7X, bract 4.9X, flower 4.1X, leaf 2.5X, seed 6.7X, legume 8.4X.

52. *Dalea candida* 0.7X, flower 4.3X, leaflets 3.2X, bract 4.9X, calyx enclosing legume 5.2X, seed 7.0X, legume 3.5X.

53. *Dalea cylindriceps* 0.7X, calyx enclosing legume 5.6X, bract 4.1X, legume 4.5X, seed 4.0X, leaflets 2.9X, flower 5.0X.

54. *Dalea enneandra* 0.7X, flower 4.4X, bract enclosing calyx 4.4X, seed 6.3X, legume 6.5X, leaf 4.3X.

55. *Dalea formosa* 0.7X, legume 4.0X, flower 4.4X, seed 2.5X, leaflet 7.0X, leaf 6.3X.

56. *Dalea lanata* 0.7X, flower 5.5X, calyx 4.9X, leaflets 4.4X, legume 7.7X, seed 6.0X.

57. *Dalea leporina* 0.7X, calyx enclosing legume 7.2X, legume 3.6X, seed 6.0X, leaflets 4.1X, bract 7.0X, flower 7.0X.

58. *Dalea multiflora* 0.7X, calyx enclosing the fruit 6.6X, bract 11.9X, legume 6.7X, flower 4.4X, leaf 3.4X.

59. *Dalea purpurea* 0.7X, calyx enclosing legume 4.8X, bract 5.7X, legume 5.2X, seed 7.0X, flower 4.1X, leaf 3.6X.

60. *Dalea villosa* 0.7X, legume 7.4X, mature calyx 5.6X, leaflets 4.8X, flower 5.0X, bract 5.5X, seed 7.0X.

61. *Desmodium canadense* 0.7X, flower 3.6X, leaflet 1.1X, loment 1.5X, inflorescence in fruit 0.7X, seed 4.2X.

62. *Desmodium canescens* 0.5X, flower 2.7X, seed 4.3X, loment 1.7X.

63. *Desmodium glutinosum* 0.6X, flower 3.8X, seed 2.9X, loment 1.5X.

64. *Desmodium illinoense* 0.7X, flower 6.1X, leaflet 1.5X, partial leaf cross-section 7.0X, seed 7.4X, loment 1.2X, inflorescence in fruit 0.7X.

65. *Desmodium nudiflorum* 0.6X, flower 3.5X, seed 3.2X, loment 1.5X.

66. *Desmodium paniculatum* 0.7X, flower 3.8X, leaflet 1.6X, loment 1.6X, seed 4.0X, panicle after shedding loments 0.7X.

67. *Desmodium sessilifolium* 0.7X, flower 6.7X, loment 2.2X, seed 3.2X, inflorescence in fruit 0.7X.

68. *Glycyrrhiza lepidota* 0.7X, flower 4.5X, leaflets 0.8X, seed 3.0X, legume 1.5X, inflorescence in fruit 0.7X.

69. *Lathyrus latifolius* 0.7X, legume 1.1X, seed 2.8X.

70. *Lathyrus polymorphus* 0.7X, leaflet 2.7X, stipule 1.9X, legume 1.2X, seed 2.9X.

71. *Lathyrus venosus* 0.7X, flower 2.4X, seed 1.4X, legume 1.0X.

72. *Lespedeza capitata* 0.7X, legume 3.2X, calyx enlosing legume 3.2X, flower 3.0X, leaflet 1.9X, seed 5.6X.

73. *Lespedeza cuneata* 0.7X, flower 5.3X, legume 4.4X, leaf 2.8X.

74. *Lespedeza stipulacea* 0.7X, legume 8.6X, flower 5.2X, leaf with stipule 2.9X, seed 6.7X.

75. *Lespedeza striata* 0.7X, legume 6.0X, flower 7.3X, leaf with stipule 2.1X, seed 4.7X.

76. *Lespedeza violacea* 0.7X, flower 6.1X, leaflets 5.1X, legume 5.0X, seed 3.7X.

77. *Lespedeza virginica* 0.7X, flower 6.1X, seed 5.0X, legume 7.3X, leaflet 1.3X.

78. *Lotus corniculatus* 0.7X, flower 2.7X, leaf 2.5X, seed 6.5X, legume 2.5X, inflorescence in fruit 0.7X.

79. **Lotus purshianus** 0.7X, flower 4.9X, leaflet 3.4X, legume 2.1X, seed 5.0X.

80. *Lupinus argenteus* 0.7X, flower 1.8X, legume 1.2X, seed 2.3X, dehisced legume 1.2X.

81. *Lupinus plattensis* 0.7X, leaflets 1.8X, seed 2.1X, legume 1.3X, flower 1.4X.

82. *Lupinus pusillus* 0.7X, flower 2.8X, legume 2.4X, seed 2.8X.

83. *Medicago lupulina* 2.7X, flower 17.2X, seed 4.2X, legume 8.7X, inflorescence in fruit 3.4X.

84. *Medicago sativa* 0.7X, flower 4.5X, leaflets 2.5X, inflorescence in fruit 2.5X, legume 3.9X, seed 3.5X.

85. *Melilotus officinalis* 0.7X, flower 4.8X, leaflet 2.3X, seed 4.6X, legume 6.7X, inflorescence in fruit 0.7X.

86. *Onobrychis viciifolia* 0.7X, flower 4.9X, wing and banner petals 4.9X, leaflet 2.0X, legume 4.1X, seed 2.4X, inflorescence in fruit 0.7X.

87. *Oxytropis campestris* 0.7X, leaflet 3.6X, seed 4.2X, flower 4.1X, legume 2.6X.

88. *Oxytropis lambertii* 0.7X, flower 2.1X, leaflet 3.4X, seed 3.9X, legume 1.9X, inflorescence in fruit 0.7X.

89. *Oxytropis multiceps* 0.7X, flower 2.5X, seed 3.9X, legume 2.2X, leaflet 2.5X.

90. *Oxytropis sericea* 0.7X, flower 2.1X, seed 3.9X, legume 2.2X.

91. *Psoralea argophylla* 0.7X, flower 4.0X, leaflet 2.3X, seed 3.1X, legume 3.2X, inflorescence in fruit 0.7X.

92. *Psoralea cuspidata* 0.7X, inflorescence in fruit 0.7X, flower 4.8X, leaflets 2.4X, seed 4.2X, legume 4.4X, calyx enclosing legume 1.8X.

93. *Psoralea digitata* 0.7X, flower 4.3X, leaflets 2.3X, seed 3.5X, legume 2.7X, calyx enclosing legume 2.2X.

94. *Psoralea esculenta* 0.7X, flower 2.3X, seed 2.7X, legume 1.8X, leaflets 0.8X, root 0.7X.

95. *Psoralea lanceolata* 0.7X, flower 5.8X, leaflets 2.4X, inflorescence in fruit 1.1X, seed 2.5X, legume 3.3X.

96. *Psoralea tenuiflora* 0.7X, flower 5.3X, seed 2.2X, legume 3.2X, leaflet 3.9X.

97. *Robinia pseudoacacia* 0.7X, flower 1.8X, seed 2.2X, legume 0.6X.

98. *Sophora nuttalliana* 0.7X, flower 2.7X, legume 1.0X, seed 3.5X, leaflets 2.0X.

99. *Strophostyles helvola* 0.7X, seed 2.9X, flower 3.4X.

100. *Strophostyles leiosperma* 0.7X, seed 4.9X, flower 5.3X.

101. *Stylosanthes biflora* 0.7X, flower 7.1X, seed 4.9X, legume 5.5X.

102. *Tephrosia virginiana* 0.6X, flower 3.2X, legume 1.5X, seed 2.1X.

103. *Thermopsis rhombifolia* 0.7X, leaflets 1.3X, seed 3.5X, legume 1.1X.

104. *Trifolium campestre* 0.7X, flower 9.0X, leaflet 4.4X.

105. *Trifolium fragiferum* 0.7X, flower 5.1X, seed 7.5X, legume 3.8X.

106. *Trifolium hybridum* 0.7X, legume 5.6X, seed 3.7X, flower 5.6X.

107. *Trifolium incarnatum* 0.7X, flower 3.6X, seed 6.7X, legume 6.0X, calyx enclosing legume 4.2X.

108. *Trifolium pratense* 0.7X, flower 5.5X, seed 7.5X, leaf 1.1X.

109. *Trifolium reflexum* 0.7X, flower 3.6X, seed 7.0X, legume 3.9X.

110. *Trifolium repens* 0.7X, flower 5.9X, seed 7.7X, legume 6.5X.

111. *Vicia americana* 0.7X, flower 3.2X, stipule 5.3X, legume 1.5X, seed 2.6X.

112. *Vicia villosa* 0.7X, legume 2.0X, seed 2.1X, flower 3.7X, leaflet 3.4X.